The Thread

INTRODUCING
THE
★AUTHOR★

The Thread

A Mathematical Yarn

PHILIP J. DAVIS

With Illustrations by Elisa M. Nazeley

BIRKHÄUSER
Boston • Basel • Stuttgart

Library of Congress Cataloging in Publication Data

Davis, Philip J., 1923–
  The thread.

  Bibliography: p. 125
  I. Title.
PN6162.D376 1982   814'.54   82-17751
ISBN-13: 978-0-8176-3097-3   e-ISBN-13: 978-1-4684-6724-6
DOI: 10.1007/978-1-4684-6724-6

CIP-Kurztitelaufnahme der Deutschen Bibliothek
Davis, Philip J.:
The Thread : a math. yarn / Philip J. Davis. With
ill. by Elisa M. Nazeley. - Boston ; Basel ;
Stuttgart : Birkhauser, 1983.
    ISBN-13: 978-0-8176-3097-3

© Birkhäuser Boston, 1983
Softcover reprint of the hardcover 1st edition 1983

ISBN-13: 978-0-8176-3097-3

To Hy and Rosalie, Tilly and Jack.
To all who joined us at the Wayside Inn.
and
To the memory of Cousin H., Master Raconteur

CONTENTS

PROLOGUE

A number of years ago in November, my brother arranged for a family party to be held in a private dining room of the Wayside Inn in Sudbury, Massachusetts. He asked me to prepare some sort of entertainment for the group.

In the week that followed his phone call, I thought about the Wayside Inn. A hundred and some odd years ago, Henry Wadsworth Longfellow had sat around the stove listening to the small talk of the landlord and travellers, and had then gone back to Cambridge to write the "Tales of a Wayside Inn."

Following the model of Chaucer's Canterbury Tales, Longfellow created his own group of travellers who came together at the Inn. With the landlord, there was the Student, the Sicilian, the Spanish Jew, the Theologian from the Banks of the Charles, the Musician, the Poet; and each, in turn, told a tale. I read the stories over: King Robert of Sicily, the saga of Olaf, the Birds of Killingworth, Paul Revere, others. Yes, this is where the famous midnight ride of Paul Revere occurs.

I wondered whether I could read some of these stories to the assembled family by way of an entertainment, a diversion during the meal. Listen, my children, and you shall hear. But would Cousin Gertie turn her ear — Gertie who couldn't sit quietly for three consecutive minutes without lighting four cigarettes and making five phone calls? Despite the sympathetic environment and the tour de force of reading Longfellow in his own setting, I decided that she would not.

The tales of Longfellow led my thoughts to Chaucer, to Froissart, the medieval French historian who studded his accounts with trumpets and peacocks and painted in colors of red and gold. I thought back to Boccaccio, to Helen of Navarre, back even to Homer: Penelope among the suitors, Nestor's sacrifice.

I thought of "Shepherd Tom" Hazard, of Portsmouth, Rhode Island, who spun a mighty yarn and whose masterpiece *The Jonny-Cake Papers* was published in his eighty-third year. "Shepherd Tom" tells how it was that the famous Rhode Island greening apple was descended from the Tree of Knowledge; how Black Phillis caused the French Revolution; how a forty pound Narragansett cheese attracted so many rats to the palace of the Nicobar Islands that the King and Queen had to jump out of the window.

Could I read this to the company? I had half a mind to, but I was beset by doubts. Still, I wanted to do something for the occasion. What?

My breakthrough came when I abandoned the idea of using other people's material. There was a line of inquiry that I had been pursuing, slowly and deliberately, for more than a decade. The time had come to tell it. The material was true. It was personal. It related in an easy-going way to my professional activity. I would put a few decorations on it and send it in to the table. What follows is the worked up version.

THE THREAD

The teller of tales speaks: the world is an infinite tapestry. There is a thread that runs between any two people, no matter when they lived or where. Find a thread and follow it: what other story do you need?

I.

TSCHEBYSCHEFF

In 1963 I brought to completion and published a book entitled *Interpolation and Approximation*. This book took six years from conception to birth. I acknowledge three of these years as my own, a legitimate period of gestation. The remaining three years were due to fouled-up circumstances. I did not, as Oscar Wilde once did, leave the manuscript absent-mindedly in a cab between Picadilly and the Strand. My publisher was anxious to keep costs down and had the galleys set up in Northern Ireland. I used to receive the galleys in batches of twenty. After several months, halfway through the book, they stopped coming; the printers were on strike.

Four months later the strike was over (an easy strike, the publisher wrote me; it was much worse in '56), and threats of international litigation over breach of contract were behind. But then my publisher was compelled to go through a financial reorganization. He was sold, or perhaps he sold himself, to a larger and more financially diverse publisher. This entailed moving his office to a different city, accumulating a new editorial and production staff, while I, a mere chattel's chattel, was left fretting and biting my nails and parrying the constant inquiry "Is it out yet?" with as much cool as I could muster. But the long sequence of postponements from "next spring" to "two more months" to "one more week" to "next Tuesday" finally and truly converged. The day arrived, and my book was in my hand.

My publisher was a bibliophile and was in the habit of having presentation copies made up for all his authors' books. I took my book out of its slip case with some care. What a handsome thing. Dark brown leather, tooled decorations, edges in marble and gold leaf. I gave it a pat and a soft kiss and placed it on my desk between two equally handsome walnut bookends that had been fashioned in the crafts program at Berea College.

One does not expect to make money from a book on advanced mathematics directed toward research mathematicians. The drug store sales are limited while the movie and television rights, though contractually assigned to the author and his heirs, are not apt to be picked up. In such circumstances one writes for reputation, for glory, one writes occasionally as a means of communication and instruction, but one does not, while in full possession of one's faculties and Dr. Sam Johnson notwithstanding, write for money. My book got some nice notices in the mathematical press in the U.S.A., Europe, the U.S.S.R. and Japan. In time, my vexation at the delay in publication abated. I even signed a contract for a second book with the same publisher. (The company that owned him had by this time itself been sold to a conglomerate and was under orders of procurators trained at the Harvard Business School who didn't care about the difference between a book and a box of corn flakes.)

But this fine reception, though greatly appreciated, was totally impersonal. There were no light-hearted telegrams at opening night, no flowers at curtain call. No flood of fan mail from the interpolators of the world inundated me. None of my colleagues ever mentioned my book in personal conversation.

About three years after *Interpolation and Approximation* appeared, I received, from a computer scientist in one of the Scottish universities, a letter that went along these lines:

"I should like to tell you how much I and my colleagues at St. Ida's are grateful to you for placing in our hands a work which is distinguished for its masterful etc. etc." (Go ahead. Praise me. I can stand it.)

"However [Ah, yes. We come now to the However. Having concluded the Gloria and the Benedictus, we come to the However, the tithe that solid thinking demands of vanity], your presentation is flawed by your insistence on spelling Chebyshev's name as 'Tschebyscheff.' This barbaric, Teutonic, non-standard orthography will gain you no friends. I sincerely hope that when you come to prepare the second edition of your book you will alter this incorrect and irritating spelling. Yours faithfully, John Begg, Professor of Mathematics."

Mathematicians are notorious hair-splitters. This is their métier. The difference between two and three can shake their cosmos. The difference between $\sqrt{2}$ and a fraction can be made as small as you please by selecting the fraction properly; yet the mathematician is prepared to sacrifice a hecatomb of oxen, as Pythagoras is reputed to have done, in asserting that the difference is eternal. This is not mere academicism; it is, for many, the whole game.

I first became aware of the mathematical temperament in action in 1944. At that time I was working for the National Advisory Committee for Aeronautics (NACA) a government research agency and the forerunner of today's space agency NASA. I had come down to the Langley Field, Virginia laboratories of the NACA together with my classmate Christopher Lockert. We had both just graduated from college, having been spared from the draft by our studies, and were assigned to the NACA at Langley by the Air Force Reserve. Lockert was brilliant; he had the kind of brilliance that penetrates intuitively and unselfconsciously. We were assigned to different sections. Lockert went to a section that concerned itself with theoretical aerodynamics while I worked in a laboratory that was measuring the loading on the tail structures of dive bombers.

At that time, a new mathematical work by the blind Russian mathematician Leonid Pontryagin had just been translated into English and had burst upon the mathematical scene with considerable éclat. Although the subject matter, the theory of topological groups, was far removed from aerodynamics, a study group was formed to read Pontryagin after hours. The war would not last forever, we figured, the cold war was unforeseeable, and we would all be wanting to beat our aerodynamic swords into more abstract plowshares. Lockert joined the study group. I did not, as I had never got much satisfaction from groups whether study or topological. Lockert got as far as page two of Pontryagin's work. On page two he found a flaw. One of the hypotheses that the author adopted in setting up the whole theory was unnecessary. Now this is largely a matter of aesthetics, but it bothered Lockert. If an author falls down badly on page two, what flaws and inelegancies, what errors, even, might be found on the two hundred pages that follow?

As Lockert told me the story, he was working overtime one evening. He was getting nowhere with his NACA problem and for a change of pace began to stew over page two of Pontryagin. Around nine o'clock he could stand it no longer and put through a long distance call to Edward Stanger who was his old mathematics professor, then working at Columbia on a defense project. In those World War II days of "Is this trip necessary?" one did not put through a long distance call lightly from an official line at an Air Force Base. The NACA switchboard was closed down and his call was routed through the Langley AFB switchboard. The corporal on duty questioned Lockert. Was this an authorized call? If so, what was the authorization number and the name of the civilian official who authorized it? Lockert said that it was an unauthorized civilian call, but that it was an emergency. The corporal responded that this was an irregular circumstance, and that he would have to obtain the O.K. of the Provost-Marshall and would Lockert explain to him the nature of the emergency. Lockert explained to the corporal as clearly as he was able that on page two of Pontryagin's book on topological groups, where the author sets up the postulates for group multiplication and group inverses, condition 2a was redundant. At the other end of the line, the corporal quickly got the point and was convinced that condition 2a was a serious one and that if it weren't checked immediately all the experimental planes would rot, their tails would fall off, their centers of gravity would be badly twisted and this condition might very well spread to the Air Force planes.

By 2300 hours it had been ascertained that the Provost-Marshall was making a night of it in Norfolk. By 2400 a Deputy-Provost-Marshall had been located and permission for the call was secured. The lines to New York were loaded, but by 0200 Edward Stanger, sleeping soundly somewhere in Manhattan, was aroused and informed of the condition prevailing on page two of Pontryagin's book.

The professor, who had put in a hitch as head proctor of Ashworth House, figured this call to be in the category of what has come to be called "the high jinks of the thirties," similar to overturned fire extinguishers or cows from Billerica farms brought up five flights to the dormitory roofs. Stanger told Lockert quite firmly to take two aspirin and write him about the condition in the morning.

Lockert never wrote. He abandoned topological groups entirely and today is one of the world experts on the theory of wave guides. Even as this drama was being acted out, somewhere in Russia, Pontryagin was also abandoning topological groups and applying his genius to the theory of stability and control which was somewhat more relevant to fighting the Great Patriotic War. Of course, we did not know about it at the time.

In the years that followed, I got used to the fact that mathematicians are notorious hair-splitters — splitters of cumin seeds, as ancient authors wrote. I adopted the stance myself; I probably promulgated it among my students. But I was never really comfortable with it. When Begg's letter came, I took it quite calmly. I had misspelled, consistently misspelled a man's name. What was so awful about that? I didn't know Begg from Beans. Was he another Seeker after Perfection? Or had his breakfast been spoiled? I felt myself on solid ground as far as his criticism was concerned. I would answer him politely, but firmly, a mere courtesy, and tell him to go fry his fish elsewhere.

I must now explain about my confidence and stubbornness. In order to do so, I must tell something about Tschebyscheff. I must explain who he was, what he was doing in my book, and why, apparently, there is some question as to the proper spelling of his name. Tschebyscheff was a famous Russian mathematician. To place him properly, I should give a brief history of Russian mathematics. But why not go back to Adam and Eve, to whom God said "Be fruitful and multiply," and place Russian mathematics properly by giving a brief history of world mathematics?

To that fictitious fellow who has developed an average sensitivity by being exposed to three or four years of high school mathematics, the subject of mathematics often appears to be static: it has always been there and it will always be there. Five plus five is ten; it always has been and always will be ten. This is far from the truth. There was a period in civilization when the concept 'ten' was first invented.

It makes a convenient though somewhat false picture to divide the history of mathematics into four periods. The Dawn starts in the cave, and proceeds through the Egyptian, Sumerian, and Babylonian periods. Then comes the Golden Age of Classical Mathematics. This starts about the time of Pythagoras (c. 550BC) and lasts through Archimedes (200BC). All is then quiet for 300 years during which time Greece became a province of Rome, Caesar was assassinated, and Christianity began to give Isis worship a run for its money.

Around 150AD, the Silver Age of Classical Mathematics set in. The stars were not quite so bright, their work was more derivative, there was a tendency to anthologize. The Silver Age lasted several hundred years. The action took place largely in Alexandria in Egypt, centering around the Museum, the Library and the Serapion (a temple dedicated to the god Serapis, a late version of Osiris). All of these are places of humanistic and philosophic paganism. These were centuries of racial and religious strife in Alexandria. The Serapion was destroyed. The mathematician Hypatia, whose father was a philosopher-priest, was slaughtered and her flesh scraped with shells. By 550AD the Academy was closed. In 638, the Calif consigned the scrolls in the Library to the furnace, to provide fuel for the public baths. (Let us console ourselves: a good fraction of these scrolls were trash — spells, enchantments, gossip, bad plays and suchlike.) Mathematics was dead; the Dark Ages had set in. The very word 'mathematicus' had come to mean an astrologer and this practice was proscribed by the laws of Diocletian and later by the Salic Laws. (I once asked Otto Neugebauer, a world-renowned authority on ancient science, why classical mathematics died. He answered me by making a coarse noise. I suppose this was his way of telling me that the answer will never be known and the question was profitless anyway.)

The Modern Period began around 1100 and its growth has been more or less continuous to the present day. It would make sense to divide this period further into the pre- and post-Newtonian periods. Influenced by the writings of the Arabic mathematicians in the Middle Ages who, in their turn, had access to certain classical manuscripts that had been spared from warming the tepidarium of Alexandria, mathematics spread through Italy, Spain, France, thence to England and Germany, thence to the rest of Europe. The mathematics developed by or available to the Englishman Thomas Hariot, the first mathematician of any originality to come to America (he came with the Raleigh expedition), was competent and copious and represents a far more substantial corpus of material than what is today taught in our high schools. It was not modesty alone that led Newton to write that he had "stood on the shoulders of giants."

The world is now in The Golden Age of Modern Mathematics. I rejoice in my contemporaneousness. Seek the reasons for the gilt if you like. It may not last. There are always baths to be warmed, and a fuel shortage is forecast.

If I have disposed of 4000 years of mathematics in a few paragraphs I must not do as little for Russian mathematics, for the particular is not included in the general. As far as we of the west are concerned, Russian mathematics begins with Leonhard Euler (1707-1783). Euler, who I believe to be the greatest mathematical genius of the past three centuries, was a native of Switzerland. (The laurel must go either to Euler or to Gauss. Euler can be compared to Mozart,

Gauss to Beethoven; so take your pick.) Off and on he worked for Frederick the Great in Germany and for Catherine the Great in St. Petersburg. Evidently, Euler did not scintillate sufficiently for Frederick, who was far more impressed by mots than by theorems. Catherine offered him patronage, Euler accepted, and wound up by being the first Russian mathematician of

The golden age of classical mathematics.

stature. Euler wrote his works in Latin, the language most frequently used in scientific writings until the nationalism of the early 1800's.

After Euler, there is a gap of a good many years before we come to a great name. We are then confronted by Lobatchewsky (1792-1856). Lobatchewsky has been twice immortalized: once for his discovery of non-Euclidean geometry, a prize which he must share with the German Karl Friedrich Gauss and with the Hungarian Janos Bolyai, and once for appearing in a ballad written by the composer-cabaret artist-mathematician Tom Lehrer. Many people who have never heard of Euclidean geometry, let alone non-Euclidean geometry, know that "Nikolai Ivanovitch Lobatchewsky is his name." Many people who think that higher mathematics consists of doing long division in one's head are familiar with Lehrer's severe judgement that the way to do mathematics is "Plagiarize. Plagiarize. Let no one else's work evade your eyes." Of course, they think that the lyricist is referring to the copying that goes on in high school examinations. So be it!

From Lobatchewsky, we move to Ostrogradsky (1801-1862). A bulldog of a man, Ostrogradsky discovered what we in the west call Stokes' Theorem. He also discovered Liouville's Theorem, with the possibility of the opposite occurrence in both instances. Russian scientific life was severely isolated from the west in the early 1800's.

Along with Ostrogradsky there is also the mathematical luminary Bunyakovsky (1804-1884). Bunyakovsky is famous in the U.S. and in England for having discovered the "Schwarz Inequality," a very useful

SCHWEIZERISCHE NATIONALBANK
BANCA NAZIUNALA SVIZRA

Euler on Swiss bank note.

theorem which, apparently, is due in essence to the French mathematician Cauchy (1789-1867). We should never have known about the true pedigree of this theorem if it had not been for the efforts of J. Stalin, né Josef Vissarionovitch Dzhugashvili, ex-theological student and for many years Dictator of All the Russias.

In pre-W.W. II days, the phrase "Schwarz Inequality" seemed to do the trick quite nicely. After the war, the word got around that it was more polite to write "The Schwarz-Bunyakovsky Inequality," though there was a feeling that mathematical authors who used this term were soft on communism. What had occurred in the interim was this: the Cold War, Stalin's paranoia, his repressive measures, and the fostering of an excessive degree of nationalism. Russian scientists who had frequently published their findings in German or French or occasionally English so that they would be more accessible to their western colleagues suddenly

Nikolai Ivanovytch Lobatchewsky

Mikhail Vasilevich
Ostrogradsky

Viktor Yakovlevich
Bunyakovsky

stopped using these languages. As Solzhenitzyn puts it, Stalin said: "Stop the toadying to foreign cultures." Only Russian and some of the minority languages such as Ukranian and Georgian were kosher. These were the pre-Sputnik years, when the Russian press put in claims for the discovery of the airplane, baseball, the zipper, and the Schwarz Inequality.

Stalin was an intellectual, let no one gainsay that, a prime example of the fact that intellect, like art, can be morally neutral. Stalin read novels, poetry; he poked around in science and in art. He messed with genetics, and deriving all conclusions from the postulates of dialectical materialism, put his blessing on Lysenko-ism. He messed with linguistics. There is no reason why he might not have browsed through a mathematics book. Some authorities say he did, but couldn't find anything to contribute. Why write "Schwarz" and risk suspicion of being soft on the west when "Bunyakovsky" would do just as well, even better?

I have deliberately exaggerated here. Many of the claims were historically just. The scientific journals where they were first reported are often obscure and not easily obtained. But once read, the claims are borne out. Each nation lives with its own scientific myths and heroes to a degree that is unappreciated. An American schoolboy and a French schoolboy will come up with a totally different set of names in discussing the inventions of the industrial revolution.

I come at last to Pafnuty Lvovitch Tschebyscheff (1821-1894). When I speak of Tschebyscheff I do so in hushed tones and I try to damp down my natural levity. This man is one of my patron saints. In fact, he is one of the patron saints of Russian mathematics, and

Pafnuty Lvovitch Tschebyscheff

it is due in no small measure to him that Russian mathematics today stands second to none in the world.

Tschebyscheff's life was his work; his vital statistics are easily recounted. He was born in 1821 in Borovsk, government of Kaluga, about 50 miles south of Moscow. His parents were of the minor nobility. In 1837 he entered the University of Moscow and graduated in 1841. He wrote a dissertation entitled "On the numerical solution of algebraic equations of higher degree", which won him a gold medal. (One hundred and thirty years and a half dozen generations of computing machines later, this topic is still mathematically active.) In 1847 he was teaching in the University of St. Petersburg and from that time to his death in 1894 he was essentially in St. Petersburg, either at the

University or at the Academy of Sciences. He spent most of his summers abroad; Paris particularly appealed to him. In Petersburg he led a restricted existence, wrapped up in himself and his work; in Paris he seemed a bit more expansive. He never married.

Photographs of him taken later in life show him to be a gaunt man with a full white Santa Claus beard and soft eyes. From other pictures in his biographies, those of mamma and papa, grandpapa, the house in Borovsk, the diplomas and army certificates, the grave-stone, the historical landmark bearing the legend: *Here was born Pafnuty Lvovitch Tschebyscheff, famous mathematician and Academician*, there emerges the distinct feeling of déja vu, something out of Chekhov, but a generation earlier. There is one vital distinction: we are not dealing here with the unmarried uncle, the ineffective failure, the gambler, the womanizer, the sponge; we have here the great professor, the Academician, the first Russian since the time of Peter the Great to be elected to the Paris Academy.

Tchebycheff's portrait on a stamp.

II.

CYRIL

I received my Ph. D. in Mathematics in 1950. As Mathematics is a vast field of inquiry the most vigorous of whose practitioners can muster perhaps only a 10% knowledge, it is worth specifying further that my thesis was written in a subfield known as "analysis" and in a sub-subfield known as "interpolation and approximation." This narrow corner of mathematics has to do, roughly, with how and in what manner curves and surfaces may be approximated or reproduced by other curves or surfaces that are mathematically simpler. For example, the shape of a circle is reproduced very nearly by replacing it by an inscribed or circumscribed regular polygon. If you replace a circle of radius of about an inch by a polygon of about fifty sides and stand the figure off about one or two feet and look at it, your eye cannot distinguish the difference between the polygon and the circle. This is the principle upon which the CALCOMP plotter, a contemporary computer-driven drawing instrument, is based. The plotter does not draw curves; it draws hundreds of tiny straight line segments, which is easier to do mechanically, and in this way, a visually acceptable approximation of a curve is built up.

This trick was known and exploited by the Greek mathematicians of the Golden Age. Since classic times approximation theory has grown in scope and subtlety so that it, too, can profitably be subdivided into three or four layers, or substrata. In 1950 there were two major centers of research in the theory of interpolation: the U.S.A. and the U.S.S.R. In the case of Russia, this was, as we shall see, a direct inheritance from Tschebyscheff. The field at that time, though alive and well, was not remarkably active. In the years when I was doing my thesis, the first generation of electronic computers had just been built and were being exploited. It was pretty clear that the mere existence of these machines was going to give a terrific shot in the arm to the subject of approximation theory, for the art of making advanced mathematical computations stands in an intimate relationship to approximation theory.

It was clear to me that I would benefit professionally if I learned to read Russian, or at least mathematical Russian. I did not want to take a regular course and learn how to say that the pen of Auntie Masha was in the garden of the dacha. I plunged in by buying a Russian dictionary and systematically translating Gontscharoff's text on interpolation and approximation. This was the cold turkey treatment and pretty painful at first, but it worked. After six months' exposure, I felt sufficiently competent to review articles in Russian for the *Mathematical Reviews*. I studied nothing systematically, least of all grammar. What makes the method possible is that in a field as specialized as mathematics, a limited vocabulary of, say, 2000 words will get you through 90% of the ideas, and anyway you kind of know what the author is talking about before you start translating. Mathematical symbols and equations are international and Auntie Masha does not normally turn up bearing circles and straight line segments.

Nonetheless, Russian is a difficult language and its difficulties begin with the alphabet.

My alphabetic troubles with John Begg can be traced to the middle of the Ninth century. Around 860, Prince Rostislav of Moravia sent a plea to the church authorities in Constantinople for teachers who would be able to instruct the populace in their own language. A mission was formed headed by two brothers from Salonika, Cyril and Methodius. These men were linguists and scholars. Cyril devised a new alphabet, called glagolitic, based somewhat on the Greek alphabet, and translated church literature into it. This is the first known instance of Slavic writing (Old Church Slavonic). The alphabets used today in Russian and other slavic languages evolved from this alphabet and are known as Cyrillic.

The Russian of the "new orthography" that came into use after the Revolution has 32 letters. Twelve of these letters are identical in shape with the Roman letters. Of these twelve, only four have approximately the same value as they do in English. Of the twenty letters that have different shapes, nine represent sounds which are not usual in English or sounds which when they occur in English would normally be expressed by groups of letters. Two letters are hard and soft signs. For example, in Russian there are single letters of the alphabet for the sounds *ch* as in cheese, *sh* as in short, *sch* as in mischief, *z* as in azure, *ts* as in tse-tse, *ch* as in the Scottish loch, *ya* as at the end of intelligentsia. Russian words and names obviously have a fixed or preferred spelling in Cyrillic. These words or names are used from time to time by English authors. How shall they be spelled in English? This is the problem of transliteration. There is no question as to the Russian spelling of Pafnuty Lvovitch Tschebyscheff. It is **ПАФНУТИИ ЛЬВОВИЧ ЧЕБЫШЕВ**

Problems of translation from one language to another are often discussed from the literary point of view. Sayings, even, have arisen on the topic: "The translator is a traitor." "To read the mother tongue in translation is like a groom kissing his bride through her veil." The problems of transliteration are of a different nature, more technical, less literary, but not altogether trivial. Transliteration might affect, for example, where a foreign author is listed in a catalog.

English typewriters do not have Cyrillic symbols nor do ordinary typesetting machines. Special symbols must be set by hand at great expense. In recent years, computer-created and -implemented fonts offer a wide variety of type faces, but the use of the original language may not be desirable because familiarity with strange alphabets is limited. What is wanted is a set of equivalents leading to expressions that are easy to read, easy to reproduce on a typewriter and which will enable the English speaker to come up with a pronunciation that approximates the original phonetically. One would like the correspondence to be fixed, and as a further requirement, a person versed in the original language should be able to retransliterate accurately. These requirements unfortunately are mutually con-

The author defends his spelling of Tschebyscheff's surname.

tradictory. All systems that have been devised are compromises.

Transliteration goes from language A to language B. But a reverse system must be devised which takes language B into language A. This may pose a different set of problems. Russian, for example, lacks the th in Timothy (the Greek theta). The Cyrillic equivalent of the Greek phi is used instead. When this name is re-transliterated into English it comes out Timofey. No great damage done here. But take a look at what happens when the name of the famous German mathematician Hilbert goes into the Russian. Russian lacks the h sound. It is usually transliterated as Г (the g sound) or as X (the kh sound). "Hilbert" transliterated into Russian and then back to English comes out "Gilbert." Mutations of this sort are sufficiently frequent in the international exchange of scientific information to cause distress among the uninitiated.

It is now time to answer Professor John Begg. "Dear Professor Begg. Thank you for your very generous evaluation of my work. Here is my defense of my spelling of Tschebyscheff's surname. In the first place, please be informed that Tschebyscheff himself often wrote scientific articles in languages other than Russian. The nationalism of the tsars (czars?) was not too severe. In his papers, his name appears in six different spellings. There is some evidence pointing to 'Tchebicheff' as his favorite spelling in Roman letters. In the second place, there is no such thing as a standard system for transliteration. There are many standards. The Library of Congress writes Chebyshev. The *Applied Mechanics Reviews* writes Chebishev. Other standards have led to Čebysev, Tschebyshev, etc. etc. But in the final analysis, it comes down to this: The spelling 'Tschebyscheff' was the one I learned in my mathematical cradle. It was there with the bottle and the rattle, and I can no more think of dropping it in favor of Čebysev than I could think of spelling the composer Tschaikowsky as Chaikovskii or Čaykovskiy.

"And now, my dear Professor Begg, before I knock off, let me give you a wad of tobacco to chew on. Why do you concentrate on his last name? Why don't you do something about his first name and his patronymic? As you well must know, our hero's full name is Pafnuty

Lvovitch Tschebyscheff. Some authorities write 'Pafnutii," some write 'Paphnutij.' Some write 'L'vovitsch.' Some write 'L'vovič'. Most mathematicians avoid the issue by not mentioning the fellow's first two names at all. For shame. How do you, sir, write? Speak and be damned. Yours in Tschebyscheff."

Begg never answered me. I figured that all orthographical questions in my book were now laid to rest.

This was too much to expect from the mathematical community. Two months after I dropped this answer into the mail box, I received a letter from the University of Bald Pate, Bald Pate, Montana. It went along these lines: "We have been using your book in our course Mathematics 345 and have found it to be of superior [etc. etc. etc.]. However, on page 84 you refer to 'Voronovsky's Theorem.' Now the Voronovsky in question is a woman. The Russians decline surnames, putting them in their proper gender and case. In future editions of your otherwise excellent work you should change this reference to 'Voronovskaya's Theorem.' Cordially yours, Charles Dickens Tierenpfeffer, Professor of Applied Mathematics."

Well, how do you like that? These transliteration fellows obviously have a club. Of course I was aware that Voronovsky (Voronovskaya) is a woman. My choice of spelling had been deliberate. I argued that in translation an author has the right to Anglicize whatever he likes. Think, for example, of Peter Ilyitch Tschaikowsky (Pyotr Ilyisch Caikoffskii). Pyotr's father was Mr. Tschaikowsky. His mother was obviously Mrs. Tschaikowsky and not Mrs. Tschaikowskaya. Nor was his mother Madame Tschaikowsky. 'Madame' is not a Russian word though it is used to lend flavor to ballet instructors and to curiosities such as Madame Blavatsky. Not to Anglicize opens up a hornets' nest of trouble. Tschaikowskaya is the nominative feminine. Why not also use the genitive, dative, accusative and instrumental forms depending upon what Mr. or Mrs. Tschaikowsky happen to be doing in your sentence?

I answered as follows. "Dear Professor Tierenpfeffer. Your point is very well taken. Many English speakers are unaware of the fact that 'aya' is the feminine Russian ending and this often leads them to commit such sexual atrocities as 'When Voronovskaya was in his 28th year.' However, I am adamant. My position is never to decline a Russian woman. Yours very sincerely."

III.

WATT

In the summer of 1967, I was having lunch with Professor and Mrs. Alexander Ostrowski on the outdoor porch of their villa in Montagnola overlooking Lake Lugano. Ostrowski, one of the world's finest mathematicians, was then professor emeritus of the University of Basel. We had first met in Washington in 1952 and had kept up a friendship. I recall we were drinking a delicious non-alcoholic wine (grape juice, really) called Traubensaft, common enough on the continent but obtainable in the U.S.A. only in specialty shops. Traubensaft is reasonably sweet and my glass was soon attacked by a swarm of yellowjackets. Mrs. Ostrowski observed my distress and knew a solution. She put a little bit of juice into a shallow dish and set it out as a decoy. The yellow-jackets were attracted to the dish away from my glass. I had never seen this trick before and I suggested to Mrs. Ostrowski that she had psychoanalyzed the bugs and knew what made them tick (she was a practicing analyst). She answered only that one must always allow nature to have its tithe.

The conversation turned to philosophy and through philosophy to Boris Pasternak who was then very much in the news. Ostrowski reminisced that he and Paster-

nak had both been students in the University of
Marburg, he in the faculty of mathematics and
Pasternak in the faculty of philosophy. The flow of
reminiscences continued and then Ostrowski remarked
that when he came up for his doctoral examinations, it
was assumed, quite as a matter of course, that every
candidate knew essentially all the mathematics then in
existence. I expressed incredulity. Ostrowski affirmed
this was so, but that it would have been the last year in
which it was so. The subject had grown so rapidly in
his lifetime that it was now beyond the power of any
one man to grasp it all. I pumped him a bit further and
asked him what fraction of mathematics he now
thought he knew. He answered that the subject could
now be subdivided into about a hundred specialties and
if one knew two or three of them one was doing quite
well.

Not all the great mathematicians of the past have
contributed substantially to all mathematical specialties. According to one historian of the subject the last
universalist was Henry Poincaré (1854-1912). Tschebyscheff was, without a doubt, a universalist of the previous generation. He worked in the theory of numbers,
the theory of probabilities, theory of integration, and
numerical analysis. He can be regarded as the founder
of the modern theory of interpolation and approximation, and as this is my principal specialty, I should like
to tell how it came about.

The subject today is of prime importance in the
devising of efficient methods of calculation for the
higher mathematical processes. Tschebyscheff, surprisingly, came to it not through problems of calculation but through the steam engine! The first steam

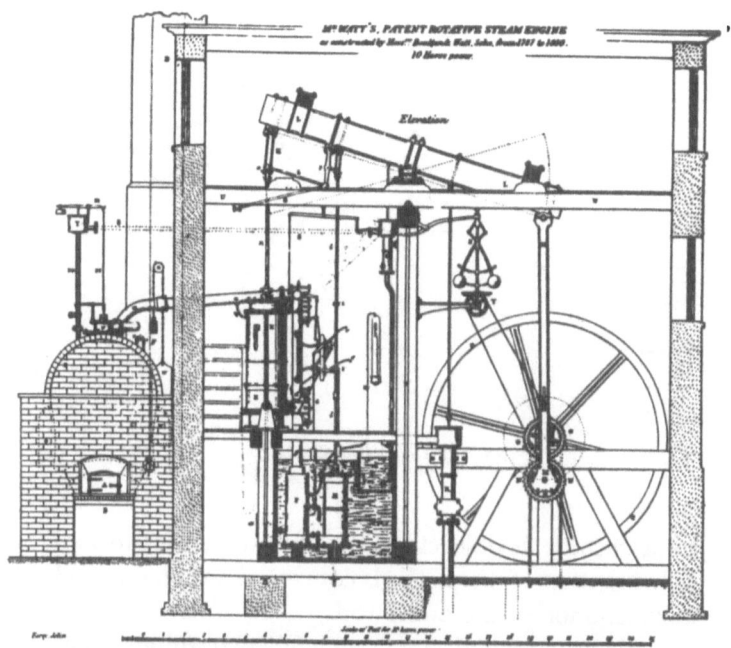

Watt's double-acting rotative engine, 1787-1800. From Farey's *Steam Engine*, 1827.

engines in wide use were those of Newcomen. They date from 1712 and were applied to pumping operations in the Cornish tin mines. Though very inefficient, they spread over England and throughout the continent. Steam engine technology did not improve until 1765 when James Watt, an obscure mathematical instrument maker working at the University of Glasgow, invented the separate condenser. In Watt's "single acting" pumping engine, the piston rod is connected to a rocking beam by means of a flexible chain. The other end of the rocking beam is connected to the pump rod

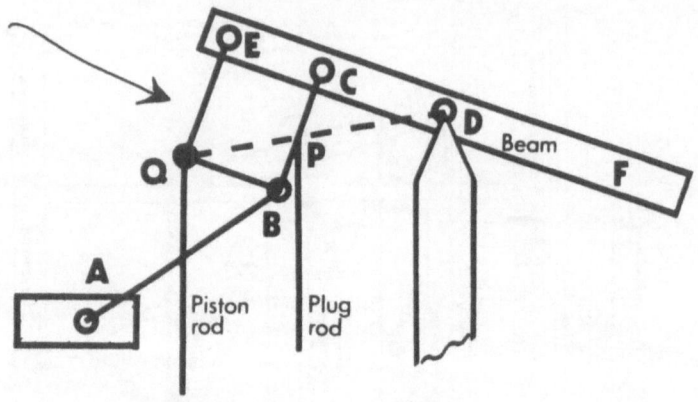

Watt's parallelogram.

by a second flexible chain. In this way, the up and down motion of the piston is converted into the up and down motion of the pump rod.

In Watt's second improvement, two things were accomplished; the single action wherein the steam pressure acts on the piston during the down stroke is replaced by the more efficient double action wherein pressure is acting through both the down and the up strokes. In addition, Watt converted the reciprocating motion of the piston rod into the rotational motion of a fly wheel. The rotational motion was what is wanted to run factory machinery. The flexible chains could not be used because although they can transmit a pull they could not transmit the push of the double action. Watt had somehow to connect the piston rod with the end of the rocking beam by a rigid connection. Now here is an incompatibility: the end of the piston rod moves mathematically in a straight line. How can linear and circular motions be combined by a rigid device?

Watt's solution was to build onto the end of the rocking beam a linkage or a pantograph-like structure of pivoted parallel rods. The end of the piston rod is now not connected to the end of the rocking beam but to one of the joints of the parallel rods. Watt, who was responsible for many mechanical devices, said that he was more proud of his parallel motion than of any other mechanical invention he had made.

Watt's device was a beautiful advance, but it was not mathematically perfect. It did not, strictly speaking, link rigid circular motion with perfect straight line motion; it did so only approximately. The difference had to be taken up by slack. With a beam about fifteen feet long and a piston stroke of four feet, there was a lateral movement of the piston of about a tenth of an inch and this resulted in a certain amount of leakage and frictional wear. (To judge this discrepancy, if an automobile piston has a separation of more than 3/1,000 of an inch between the piston and the piston rings, the car will "burn oil" and it is time for a ring job or reboring the cylinder.)

From his early childhood, Tschebyscheff loved to play with mechanisms. In later life he devised mechanisms or linkages for a wheel chair, a row boat, and a mechanical calculator which is still on display in the Conservatoire des Arts et Métiers in Paris. In the summer of 1852, he undertook a grand tour of Europe. He visited as many of the major figures in the world of mathematics as he could arrange; he also made tours of inspection of various types of machinery, windmills, water turbines, railways, and iron works. He got interested in Watt's parallel motion. The piston in Watt's device has zero lateral discrepancy at three points in its cycle. It occurred to Tschebyscheff that a somewhat different arrangement of bars would lead to a discrepancy of half of Watt's and would be zero at five points in the piston cycle.

At this point, Tschebyscheff's dominant mathematical imagination took over from his mechanical imagination, and starting from this observation he wrote a paper, now considered a mathematical classic, which laid the foundations for the topic of best approximation of functions by means of polynomials. Tschebyscheff never went back to steam engines; his suggested improvement, as is so often the case, was only theoretical; if carried out, it would have involved more moving parts with their own problems of inefficiency.

If one can get a linkage which is an improvement over Watt's, would it be mathematically possible to devise a linkage which converts straight line motion to circular motion without any discrepancy whatever?

In the 1850's this was an open question. Tschebyscheff thought so but wasn't certain.

IV.

LIPKIN

When two streams of activity fuse into one another, each of which is perfectly familiar in its own right but appears to be totally unrelated to the other, then on the day-to-day scene we are apt to call it a coincidence; on the mathematical scene it becomes a theorem.

The young Tschebyscheff concerned himself with such a mathematical coincidence and his accomplishment was such that if he had done nothing else, it would have secured him an international reputation as a great mathematician. The two streams of activity have to do with the prime numbers and the logarithms. The prime numbers are numbers such as 5, 7 or 11 which cannot be expressed as the product of two other numbers. Thus, 14 is not a prime number because 14 = 2 × 7. All numbers can be decomposed into the product of prime numbers, as for example, 18 = 2 × 3 × 3, so that the prime numbers can be considered to be the building blocks of all the numbers as far as multiplication is concerned. Despite this fundamental property, the prime numbers exhibit certain unexpected peculiarities. There are no simple formulas for finding prime numbers or for testing whether a given number is or is not a prime. There are an infinite number of prime numbers (a fact known to the ancient Greeks; a proof of this was given by Euclid in his Elements). They become more and more rare as one considers larger and larger numbers.

The distribution of the prime numbers is peculiar. They seem to pop up in odd places. Thus, the only primes in the 900's are 907, 911, 919, 929, 937, 941, 947, 953, 967, 971, 977, 983, 991, and 997. What's wrong with 931? It "looks" prime enough. Well, $931 = 7 \times 7 \times 19$, so that's the end of the matter. Every once in a while two primes come together as with 11, 13 or 281, 283 or 2087, 2089. These are called twin primes and are exceedingly rare. It has been conjectured that there are infinitely many pairs of twin primes. This conjecture is unproved as of 1973[*], and whoever succeeds in proving it will earn a reputation such as Tschebyscheff earned in 1850.

Enough for the prime numbers momentarily — what about the logarithms? These are johnnies-come-lately on the mathematical scene, having been introduced in 1594 by the Scottish amateur mathematician John Napier, Baron of Merchiston. For each number n, Napier computed another number which he called the logarithm of n, log n. Log n is an infinite decimal generally rounded off to four or five places. When I was in high school, we drilled for months on logarithms, performing complicated multiplications and roots by their use. The point is, and this is Napier's discovery, that by their use, multiplication can be replaced by addition, which is a much easier operation to carry out. Today, there is no reason to drill on logarithms, for the

[*]Added in 1983: Still unproved.

omnipresence of cheap computers makes it pointless. However, the logarithm retains its importance because it turns out to be one of the principal mathematical curves or functions. It is the inverse to geometric growth and as such it is omnipresent in science.

Now what do logarithms have to do with prime numbers? Nothing, ostensibly; at least two hundred years of working with logarithms turned up no great connection, though the realization that both somehow were fundamental to multiplication might have foreshadowed such a connection.

In brief, Tschebyscheff showed that if n is a number, the number of primes less than n was equal approximately to n + log n. This approximation gets better and better as the number n gets larger and larger. Tschebyscheff's Theorem is one of the first in what has since become a vast subject called analytic number theory.

I have traced Tschebyscheff's stream a sufficient distance. I shall now ascend a minor tributary. My story turns from the University of St. Petersburg in the 1850's to the Jewish communities of Lithuania; not so many miles, perhaps, as the wagons went, but civilizations apart as far as the set of the mind was concerned. Narrow, both excluded and self-isolated, intensely religious and often intolerantly so, pietistic, brilliant, studious, persecuted, often living in fear of bodily harm or expulsion, the Lithuanian Jews lived in a world of their own which their descendants who emigrated rejected violently and the strengths of which have only now begun to be recognized.

I shall begin biblically by saying that Samuel Lipkin who was a rabbi at Plungian begat Israel. Israel Lipkin

Dr. L. Lipkin

who lived in Friedrichstadt begat Zeev Wolf who became rabbi at Telshi. Zeev Wolf begat Israel Lipkin. Israel married a woman from Salant and for a while lived in his wife's town. Rabbi Israel Salanter (as he was generally called) became a great Talmudist and the founder of a movement that stressed ethical values. His fame spread throughout the Jewish world. Israel begat

Yom Tov Lippman. Lippman Lipkin was a student of Tschebyscheff and the first to publish the details of a linkage that converted circular motion into mathematically perfect straight line motion. Lippman Lipkin died of smallpox at the age of 25. A brilliant mathematical career had hardly begun.

But let us back up a bit. Israel Salanter (Lipkin's father) was born in 1810 in the Lithuanian province of Samogitia near the Russian border. His father was his first teacher of Talmud and Rabbinic Literature. Somewhat later he was influenced by two men, Hirsch Braude and Zundel, both rabbis, but men of quite a different stamp. Braude was the supreme dialectician, Zundel the austere and ascetic saint. In his own life Salanter tried to combine these two modes by applying dialectics to ethics and morality.

If one could isolate after the manner of the physicists the principal force vectors acting on the Jewish spirit in the Lithuania of the 1840's, one would list among the internal forces Talmudism, Chassidism, and the Haskalah. Among the external forces was the ferment of the enlightenment attendant upon the opening up of the world of arts, literature, science, and commerce. There was consequent pressure to assimilate and in many instances to convert. Talmudism was intellectual, stressing the primacy of the Talmud in human affairs. That vast compilation of law, lore and custom with layers upon layers of interpretation and reinterpretation was studied with intellectual fervor and zest. Chassidism, two generations old in the youth of Salanter, was the latest manifestation of the cyclic reaction against dry-as-dust Talmudism. Essentially anti-intellectual, Chassidism stressed the heart above the mind, the spirit above the intellect, and sought to open up springs of human experience that had been bound by the straitjacket of Talmudic legalism. Simple with strong tendencies toward mysticism, superstition, miracle working and messianism, in two generations Chassidim developed strong religious oligarchies of saints and rabbinic courts of retainers, and threatened to schismatize the Jewish communities of Eastern Europe. In Chassidism, extreme piety can come full circle and almost — but not quite — join hands with license. The world of the Chassidim, in the songs of their modern troubadours Martin Buber and Isaac Bashevis Singer, is the closest that the Jews come to the rich medieval tapestries of Froissart's History.

The schism did not occur and this was the result of the countervailing forces of the Haskalah. Haskalah is modernity, secularization, the Jewish enlightenment at the hands of Moses Mendelssohn and his followers. The Haskalah was free thinking, anti-clerical, long on social action, short on faith. When, after a hundred years, it had fulfilled its historic mission and spent its energies, it emerged revitalized in the 1880's in the new garments of Socialism and Zionism.

The principal dangers of the Haskalah, as the orthodox establishment assessed the situation, was that the Maskilim (those who embraced the enlightenment) were the tools of the government who wished consciously and unconsciously to deJudaize the Jews. There was considerable truth in this. Conversion, it was found, often opened many doors, social, professional. In the Universities, with at best very few jobs available, the pressure was severe: convert and advance. Recall that even in liberal England, professors had to swear by the 39 articles of faith of the

Church of England well into the 19th Century. The number of conversions of Jews mounted. It has been estimated that a minimum of 100,000 conversion ceremonies were performed in the 19th Century. Some authorities think a quarter of a million is a better figure. Convert, forget, pass, and be accepted.

Israel Salanter's son, Yom Tov Lippman Lipkin, was born in 1851. The language spoken at his cradle was Yiddish, and at an early age he was introduced to the intricacies of the Talmud. One may assume that with his increasing reputation and manifold activities, Israel did not hold the reins too tightly, for in adolescence his son fell under the influence of the Maskilim. He was a brilliant student. He had a natural tendency for mathematics and he taught himself as much as he could from the scanty scientific literature in Yiddish and Hebrew. He swallowed all that was available but his thirst was not quenched. He needed more. He taught himself a sufficiency of German and French, and at the age of 15 left his father's home and went to the university to study mathematics. He did not return.

Lippman Lipkin first went to the University of Königsberg. Through the influence of a certain Professor Rischelo who was convinced of the boy's brilliance, he was admitted to the lectures. After some time in Königsberg, he went to the Gewerbe-Akademie in Berlin and then to the University of Jena where he received a degree. His thesis was on the topic of space strophoids and, as far as I can make out, it was unpublished.

A number of years ago there was a member of the mathematics faculty at Brown University by the name of R.C. Archibald, who was a strong mathematical bibliophile. Singlehandedly during his career at Brown, Archibald built up a mathematical archive which is one of the best in the country. I have in my hand a rare item. It is a collection of six pamphlets bound together with a 1925 correspondence between Archibald and the Librarian at the University of Jena. The pamphlets are the "Reminiscences of the Mathematical Society" at the University of Jena, published sporadically between the years 1859 and 1882. They cover the years in which the young Lipkin was doing his thesis there.

These reminiscences give an excellent picture of what the mathematical atmosphere was like in those days. The society was started in 1850 and its objective was to bring together faculty and students informally for mathematics and sociability. At each meeting a talk would be presented either by a student or by a faculty member, and running my eye down the lists of talks, I can see that they run a wide gamut from topics that were currently on the research scene to the perennial chestnuts.

The club got off to a slow start; it seemed hard to get it going again after the sleepy summer months. After a half dozen years it was on a firm basis. A motto for the club was proposed: "More than the truth, mathematics prefers the finding of the truth," an idea which was popularized by Lessing. The method, the process; in today's lingo, the medium is the message. The motto was accepted. It was written at the top of a board and when members submitted solutions to club problems, their work was tacked up under the motto.

The reminiscences contain a list of the talks, and a directory of club members. There are personal items such as "Hermann Muller has gone off to America, and we wish him all possible luck in his new life on the western plains of Chicago." There are also extensive death notices for members.

I do not find Lipkin's name among the members, nor did he present any talks. One must imagine a young man in an unfamiliar atmosphere, wearing strange clothes, possessing strange mannerisms, speaking the language imperfectly, adhering insofar as possible, if not out of conviction, then out of inertia, to the rituals of his ancestors. One can imagine him torn between a desire to participate in the Mathematics Club and the difficulty of discarding enough of his past to enable him to do so. If the Mathematics Club served refreshments at the end of its meetings, this alone might have been enough to keep him away.

In 1861 Lippman Israïlovitch attended the mathematical lectures in a German university. Lippman 'Sruels might have sat in a yeshiva in Lithuania studying Talmud. How are they to be distinguished? Are not both Mathematics and Talmud bodies of material with roots in antiquity? Are they not both intellectual pursuits? Does not each have a characteristic method and a core of hypotheses from which one moves inevitably to the conclusions? Is not each a vast structure put together over the centuries by the work of the devoted mediocrities and enlightened and recharged by the insights of the few who were geniuses? Does not each draw from the affairs of the universe, or if one wills, cannot each be locked in sterile rotations of thought with the brain feeding merely upon the brain? Did not Elijah of Vilna, that great Talmudist, say that if an angel offered to reveal to him the whole Talmud in a flash he would reject the offer? That is for the world to come, he said. Only the things which are achieved with difficulty and great labor have any value in this world. Did not, in direct parallel, the Mathematics Club in Jena put up as its motto: More than the truth is the search for the truth? Can not one walk away from both with the one word verdict: "Irrelevant"?

History is always written from the point of view of the sons, never the fathers, for time moves in the direction of the sons. When Mr. Handel came down hard on little George Friedrich, we say of the father, what a stubborn, stupid, old fool. When we read that clever Oliver Wendell Holmes, Sr., the doctor, told his son, Oliver Wendell Holmes, Jr., the lawyer, that he would never be able to make a mark in the law, we are left feeling how wrong fathers can be. Of course, these are cases in which the sons were right and the fathers wrong. Where, as in the story of Lipkin, death left the outcome ambiguous, we can sympathize somewhat more with the father.

Word got around of the brilliant career that Israel Salanter's son was making for himself in university life. *Ha Maggid*, one of the first newspapers published in Hebrew and the voice of enlightened orthodoxy, carried a story on it. It came to the attention of the father. He felt that his public position demanded a response. "I stand at the gate of the city," he wrote to *Ha Maggid*, "and declare that the news about my son is no great adornment to me. On the contrary, it has caused me great bitterness of spirit. My heart aches at the path my son has chosen to tread. Whoever loves him and knows how to speak to the heart of my son to turn his will that he not leap off contrary to my desire

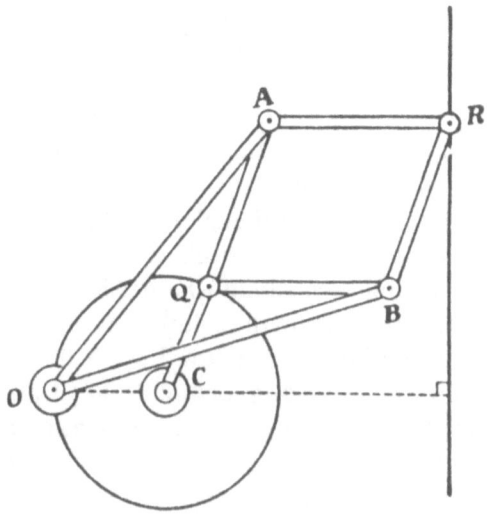

Peaucellier's Inversor. Circular motion at Q is converted to linear motion at R.

will do a great favor to one who this day is as smitten in spirit as I."

Bitter? Perhaps; but not embittered. When Lippman was in Petersburg, his father made a special trip to visit him. He wanted to exact from him the promise of three things. The first was that he keep the Sabbath. The second, that he keep the ritual laws of diet. The third, that he not shave his beard.

Lipkin died in 1875. His doctoral thesis was nearly complete. Perhaps the manuscript can be found in the archives of the university in Leningrad. Perhaps his ideas were extracted by contemporaries and have, in fact, entered the mainstream of mathematics. I do not know.

So Lipkin solved the mathematical problem of converting circular motion into straight line motion, a problem for which Watt's parallelogram was only an approximate solution. Why then, does Lipkin's name gather dust in the roll call of mathematicians? For at least three good reasons.

In the first place, even though the solution had been unknown to the great Tschebyscheff, it turned out to be too easy. Any high school kid who has had one year of geometry can understand the solution. Nothing can kill a problem so rapidly as triviality. I speak from experience. When I was working for my Ph.D., I asked my thesis adviser for a thesis topic. He suggested a problem, one that was unsolved and he thought was difficult. I took it home, and solved it in a few days and in about a paragraph. Bye bye thesis topic.

In the second place, the solution was of very little importance, technologically speaking. Lipkin's linkage required seven bars and six joints, and in practice it turned out to be less accurate than previous mechanisms because of the large number of errors due to machine tolerances.

In the third place, by one of these confluences in human affairs which appear strange but are really not so rare, Lipkin had been scooped. In 1864, a Captain in the French Corps of Engineers by the name of Peaucellier, who was stationed at Nice wrote a letter to the *Nouvelles Annales de Mathématiques* in which he claimed that he had worked out a linkage that did the job. He gave no details. Lipkin's work provided the first details in print. In 1873, Peaucellier published a solution identical to Lipkin's and claimed priority for it. The question of priority gave rise to a considerable controversy between the French School and the Russian School. Reading the details of this controversy one hundred years later, it strikes me as sterile, because of reasons one and two. In the English literature, the mechanism goes by Peaucellier's name.

I have told this story at some length because I put it together over a period of years, like the pieces of a jigsaw puzzle. I heard about Peaucellier in high school. I was first introduced to Israel Salanter through the writings of my wife's grandfather, who had been a student of a disciple of Salanter. The matter rested there for many years. In the fall of 1972, reading the Inaugural Lecture of Professor Talbot of the University of Lancaster, England, I was made aware of Lipkin, but it took me several more months before I made the connection between Lipkin, Salanter, Peaucellier, and Tschebyscheff. This is my theorem. There are thousands like it. They lie around unrecognized.

V.

PAFNUTY

I have already told how I answered John Begg's criticism of my spelling of Tschebyscheff. My reply must have satisfied him or put him off, for it terminated the correspondence. In my answer, I alluded to a similar problem of transliteration that exists for Tschebyscheff's first name and his patronymic. In the weeks that followed, I kept thinking about Tschebyscheff's first name: Pafnuty. Pafnuty, Pafnuty, Pafnuty. Curious name, is it not? It sounds peculiar. In American ears it sounds almost silly. There are not too many Pafnutys around in the telephone book. Not too many around in the Moscow telephone book, I'd wager, assuming there were a telephone book in Moscow, which until recently there was not. I was fairly familiar with Russian names. The Vladimirs, Pavels, Pyotrs, Dmitris, Sergeis, Vassilys, the Nikolais posed no problems for me. I could even manage with a Vyacheslav, Myacheslav, Svyateslav, or Vsyevolod. The Igors, Alexanders, Leonids, Timofeys were duck soup. But Pafnuty? What have we here? What kind of a name was Pafnuty? Pafnuty had me stopped cold.

There are experts on everything. There are experts on the manufacture of nylon stockings and experts on local town histories in Tasmania. There are experts on famous horses of the past and on trade journals for burlap bag salesmanship. There are obviously experts on Pafnuty. The problem as I saw it was not to consult an expert, but to arrive at the truth at my own pace in my own bumbling way. Not the truth, but the search for the truth; the process, the method, that is what matters.

I was stopped cold and things stood that way for five years. Then suddenly, quite unexpectedly, I picked up a thread, and the Tschebyscheff file was reopened.

Professors are nothing if not bookish. One Friday afternoon late in the fall I found myself passing through the stacks of the Rockefeller Library at Brown. I must bring home some light stuff for the weekend, I remember thinking. I was then walking through Religion. I felt that religion was unlikely competition for a good detective story, but reached in anyway and pulled off the shelf the first likely book I saw: "Ecclesiastical History" by one Socrates Scholasticus. Well, why not? One man's specialty may be the next man's bagatelle.

Saturday morning I sat down in front of a fine dish of eggs and anchovies. I found the combination so exquisite that I bethought me of Rev. Sydney Smith's remark that heaven was paté de fois gras eaten to the sound of trumpets. In this way I reminded myself of Socrates' book still sitting in my bag; whereupon I fetched it and propped it open on the coffee can. Why read the back of a coffee can when at no extra cost one can read the Church Fathers? Here is what stared me in the face:

Book I, Chapter XI: "Of the Bishop Paphnutius."
Again, a double take. Paphnutius, Paphnuty, Pafnuty,
Paphnutius, Paphnuty. Aha! The same name, un-
doubtedly, with a Russian ending.

Socrates Scholasticus, a Constantinople lawyer,
devoted a good part of his career to writing Church
History (he was born around 380 A.D.), and he tells of
a certain Paphnutius who was bishop of one of the cities
of upper Thebes (around 325 A.D.). The bishop was a
man of great piety and extraordinary miracles were
performed by him. In the time of the persecution, he
had had one of his eyes put out and had been ham-
strung. But the Emperor Constantine was himself so
devout that he honored Paphnutius exceedingly and
kissed his blank eye. Some authorities wrote that Constantine would touch his own eye to Paphnutius' blank
eye and surround his limbs with the royal purple; by
this symbolic translocation of the royal flesh he was
trying to compensate for what the saint had suffered at
the hands of his persecutors.

I read on. It's interesting stuff. Those were the days
when the fundamental dogmas and practices of the
Catholic Church were being forged. The synod at
Nicea (325 A.D.) promulgated the Nicean Creed. The
opinions of Arius were declared blasphemous and were
anathematized. What were these "contagious and pesti-
lential" opinions? That "the Son of God sprang from
nothing." That "there was a time when he was not."
That "the Son of God was possessed of free will, so as to
be capable of vice or virtue."

One of the practices being thrashed out related to the
celibacy of the clergy. A group of bishops made a
motion to the effect that all men in holy orders,

bishops, presbyters, deacons, should thenceforward have no further conjugal relationships with the wives they had married prior to ordination. Paphnutius rose in the middle of the assembly of bishops and spoke eloquently against this motion. "Marriage is an honorable estate and the nuptial bed is undefiled. Not all men can bear the practice of rigid continence [the Greek word for continence, transliterated, is apathy], nor would the chastity of their wives be preserved." Good man Paphnutius, himself chaste among the chaste, continent among the continent (for he had been brought up from a boy in a monastery), to have spoken thus. We should be hearing about him in the newspapers in view of the current flap on priestly celibacy. What happened back in Nicea? Paphnutius' oration killed the motion and the priests were allowed to do as each saw fit.

Madness leads men to scale mountains simply because they are there. This old story told by Socrates Scholasticus stoked up my fires to track down all the Pafnutys of the world. Now that I had made some progress on my own, I could approach the experts and not feel so nakedly uninformed. I first made overtures with Professor William Carpenter of the Department of The History of Mathematics at Brown. Carpenter is one of the great cuneiformists in the world, a scholar of ancient history, languages and science. Over the phone I explained my problem to him, telling him of my deep professional love for Tschebyscheff, and my concern for the spellings and the origins of his name, particularly his first name. I told him of my theory that the Russian Pafnuty was derived from the Greek Paphnutius, that there had been a bishop of that name, famous in church history, who had lived in Egypt.

I told him also of a further theory. Though my knowledge of Greek was miniscule, the name Paphnutius did not ring right in my Greek ears. I was familiar with Archimedes, Socrates, Aristophanes, Pericles, an occasional Pappus or Eudoxus. Paphnutius didn't seem to be one of that crowd. I figured that it was an Egyptian name. I wondered whether there was a Pafnuty among the Pharoahs. I wondered whether there were Pafnutys devising linkages for hauling stones up the pyramids.

Carpenter listened to my theory sympathetically. He told me with the surgical precision of a specialist that I had better talk it over with his colleague José Ruiz of the Egyptology Department. I told him that I didn't know Ruiz and that I didn't even know that Brown had an Egyptology Department. He answered that it was the only one in the country and a damn good one too. "How come a place as small as Brown has an Egyptology Department?" I asked him. "It's much too complicated a story to tell you over the phone, but if you meet me in five minutes, I'll tell you over a cup of coffee." Then Carpenter added, "It's all because of Boss Tweed," and hung up.

VI.

THEODORA

"*Between the years 1868* and 1871," said William Carpenter, "Boss Tweed and his cronies managed to steal about $45,000,000 from the City of New York. Figuring an inflation ratio of 10 to 1 at the very least, you can see that this was a very fine piece of cash indeed. One of the beneficiaries of this plunder was a man by the name of Charles Edwin Wilbour. Wilbour was descended from an old Rhode Island family that had settled in Little Compton in the 1690's. He was a bright boy with an interest in languages. He entered Brown with the Class of 1854 and won a prize in Greek.

"In 1854, Wilbour set out for New York City. He landed a job with the *New York Herald* as a reporter, and as Horace Greeley took a liking to him, he rose rapidly in the world of publishing. Through these connections, Wilbour met Boss Tweed. Tweed took over the *New York Transcript* and installed Wilbour as manager. Simultaneously, Tweed bought out a printing firm called the New York Printing Company and installed Wilbour as its president. The *Transcript* became the official newspaper of the City of New York and all municipal advertising was channeled to it. In 1871 Tweed gave orders to the Board of Education that all textbook bids from Harper's were to be rejected because Harper's *Weekly* was publishing the anti-Tweed cartoons of Thomas Nast. He also gave instructions that all textbooks then in use in the school system and published by Harper were to be destroyed and replaced by books published by Wilbour's company. In a period of thirty months the bill for public advertising came to more than $7,100,000. The bulk of this sum found its way into the pockets of the New York Printing Company and its officers.

"When the debacle came in 1871, Wilbour and his family packed up hurriedly and sailed for France. He was then thirty-eight.

"And now comes the remarkable part of the story. Having buried one career in America, Wilbour took up a second career. He became a gentleman Egyptologist. And a very good one."

"Like Schliemann and Troy?" I asked Carpenter. "Same idea. Here comes Ruiz. I'll show him where we're sitting and then get myself another cup of coffee."

He introduced me to José Ruiz of the Brown Egyptology Department. Ruiz is short and slight, an ascetic-looking man. I immediately recognized him as one of the regulars who take an afternoon constitutional in front of my house. "You mean to say that all these years a Pafnutyologist has been walking up and down in front of my house and I didn't know about it?"

"You can bother him later with your Pafnuty nuttery," said Carpenter, "I'm telling the story of Wilbour." Ruiz said that Wilbour was a very nice man indeed because Wilbour was paying his salary. "Tell him about the cucumber slicer," said Ruiz to Carpenter.

"Wilbour set up in Paris. He had entrée to the French literati and became friendly with Victor Hugo. He also met Gaston Maspero, who was one of the leading Egyptologists. His interest in the ancient Egyptian civilization was kindled or rekindled by Maspero and he studied on his own for a while. In 1880 he travelled up the Nile on a government gunboat. In 1886, he did it up brown by buying a houseboat — a dahabiyeh — and spending every winter on it.

"In the last years of his life, Wilbour sailed up and down the Nile, poked into all the temples, read the inscriptions and texts in heiroglyphic or Greek, acquired a few antiquities every now and then. In 1889, he made some interesting finds on an island in the First Cataract of the Nile. He died in 1896 and was very well respected by all the men in his field. He never published anything."

"How many men were there in the field?" I asked.

"Three, four, five, six, maybe. It depends how you count," Carpenter answered.

"Tell him about the cucumber slicer," said Ruiz.

"After Wilbour's death, Mrs. Wilbour settled in Central Park South. Ultimately, Wilbour's collection of Egyptian antiquities found their way into the Brooklyn Museum. Of Wilbour's four children, his daughter Theodora lived the longest. She appears not to have married and she spent her declining years keeping her father's memory at a high polish. There were periodic state visits of inspection to the Brooklyn Museum. She was a tall and heavy-set woman, but by that time she was in a wheel chair. The royal procession would make its way slowly down the galleries. On her right, the curator of the Museum. On her left, her lawyer.

Wilbour Hall.

Behind her, pushing the chair, a museum guard. She would be pushed past Papa's stuff; and seeing that it was in good shape and well displayed, she would nod her approval. Then she would be pushed over to take a look at the collection of English silver she had assembled and loaned to the Museum.

"One year she made the rounds with a new curator. They inspected the silver collection. It contained an 18th century cucumber slicer. The curator raised some kind of question about the slicer. Theodora took issue and an argument ensued. Before the lawyer could say the word 'codicil,' the collection was transferred to the Boston Museum of Fine Arts and the great expectations of the Brooklyn Museum were at an end. And this is where Brown comes in.

"In 1931, Theodora remembered her father's Alma Mater, although the old man never graduated from

Brown nor appeared to have taken any interest in the place while he was alive. In 1947 Theodora died, and Brown got three quarters of a million dollars specifically to support the Department of Egyptology."

One good story deserves another, I've always thought, so having heard in considerable detail the Case of the Insulted Cucumber-Slicer, subtitled, How Brown Got Three Quarters of a Million for Heiroglyphics, I launched into the problem of Pafnuty for Ruiz' sake, telling him more or less what is included in this essay up to this point.

Ruiz listened with scrupulous attention. Then we broke up as one of us had to teach a class. I left Ruiz with his assurance that he would look into the problem and let me know tomorrow.

Tomorrow did not come until three or four months later. I received a letter from Ruiz postmarked from an obscure station in Egypt which I could not decipher. His letter went along the following lines. "Dear Mr. Davis. I write to you from the banks of the Nile. I have been ascending temple walls and reading inscriptions. I have been making latex impressions before, alas, these noble buildings slip beneath the waters of the Aswan Dam. These structures are as old as anything in so-called civilization, but history is sufficiently ironic; who knows but they might rise again?

"One inscription, in particular, put me in mind of the question you raised to me and Dr. Carpenter. I'm afraid I know no Russian so I can't tell whether Pafnuty is truly Russian or not. Yet, I would guess you are right in thinking it derives from Paphnutius. I would therefore suppose that when Chebikoff was born [Ruiz had obviously forgotten Tschebyscheff's name], the father Chebikoff rushed to the nearest calendar of saints and selected a name for the baby who was to become so famous a mathematician.

"As for the name Paphnutius, your guess was partially correct. It is not Greek, but Coptic. You probably know that Coptic is the latest development of the old Egyptian language. The name is a variant of Papnute. There are several other forms of the same name. Its literal meaning is 'the man of God' or 'the one belonging to God.' It is not attested to in Pharoanic Egypt, but was used as a personal name among the Christian descendants of the ancient Egyptians — the so-called Copts. There were numerous ascetics, anchorites by this name. There are at least two saints by this name.

"There was Paphnutius the Bishop in Sais in the Delta. There was Paphnutius the Simple who lived in the monastery at Scaetia and was so terribly shy of women. Then there was Paphnutius the Gardener whose spiritual excellence was so great that for a period of eighty years he wore only two shirts.

"The name lingers on. Apart from your mathematician Chebikoff, there are even now many Egyptians, mostly Copts, who bear the name in the modern form of Babnuda. The pronunciation evolved from Papnute, you see.

"If you want to read more stories about different Paphnuti, read the *Vitae Patrum* which unfortunately with limited library facilities at this outpost I cannot put my hands on. Otherwise I would tell you myself. Yours sincerely."

And with that homework assignment passed out from the sands of the Nile, José Ruiz signed off.

VII.

PAPHNUTIUS

My knowledge of Latin is small. Many years ago, in Lawrence, Massachusetts, under the carrot and the whip of Susy Jordan, I got through the Gallic Wars of Caesar. Today, my Latin is limited to such matters as 'habeas corpus' and "S.P.Q.R." I see the latter deployed occasionally in movies featuring Claudette Colbert or Elizabeth Taylor.

What and where is the *Vitae Patrum* alluded to laconically in the letter from Ruiz? Now that I had found out what expertise could buy in my cause, I was no longer afraid to use it. I recalled the amo, amas, amat in Susy Jordan's class and I reminded myself that my high school classmate Phil Levine was professor of classic languages out on the coast. I didn't know where among the shifting sands of Egypt Ruiz would be, so I decided to write to Phil. He was, I had heard, a Latin paleographer of world renown, a familiar figure in the Vatican Library with students now teaching classics in universities all over the country. Never underestimate the power of a devoted teacher; her disciples are everywhere. I am talking about Susy, of course.

In my letter to Phil, I excused myself for a hiatus of thirty years during which I had been out of touch. I went on to explain my devotion to Tschebyscheff, etc. etc., and wound up by asking him what the *Vitae Patrum* was and if he wouldn't mind translating a few choice paragraphs of it for me.

Levine wrote to me almost immediately along the following lines. The quarter of a century was but a moment. The halcyon years with Susy Jordan were ever in his memory. And now to get down to business. "The Vitae Patrum are the lives and the sayings of the Desert Fathers, the Christian ascetics who lived in desert isolation. The stories were originally written in Greek but during the fourth, fifth and sixth centuries they were translated into Latin by such writers as Pelagius the Deacon and John the Subdeacon. They were published in Latin most recently in Antwerp in 1628. The collection was edited by a scholar by the name of Rosweyde and printed at the Plantin Press. The press was owned by a certain Baltazar Moret whose father-in-law, Christophe Plantin, created a number of very beautiful type faces (a fact which interested him very much as a paleographer). Rosweyde's collection is huge. It is a folio of over a thousand pages, double column. Some of the more scandalous stories (St. Simeon Stylites sitting on the top of his pillar) are well known to one and all. The book is a mine of stories, but has not been completely translated into English. There is a translation of part of the material by Helen Waddell called, *The Desert Fathers*. Start with Waddell; perhaps what you want is there. Good luck on your quest."

It was indeed. I found it in a chapter called "The History of the Monks of Egypt." The date: around 394 A.D. The author: an unknown monk, from a mon-

astery on the Mount of Olives, who made a pilgrimage
to the holy places of Egypt to visit and talk to the holy
men who lived in the desert cells.

The traveller met up with an old man by the name of
Benus. Benus was renowned for his quiet ways, his
humility, and his gentleness. The traveller asked Benus
to tell him by way of instruction what his experience
had been. Benus told him that some time back there
was a certain beast called a hippopotamus and it was
ravaging the countryside not too far away. The famers
of the district asked Benus whether he could do something about it. Benus went to the place and found the
gigantic beast. Then he said to it: "I command thee, in
the Name of Jesus Christ to lay waste this land no
more." The beast fled, "as though an angel had given
chase to it." A refractory crocodile responded to the
same treatment.

"Lay waste this land no more!"

Perhaps you don't like crocodile stories. Well, here is a horse race story with the same flavor and from the same period. It should appeal to anyone who comes from Providence, Rhode Island, which has a good half-dozen tracks within a stone's throw.* It can be found in the *Vita Sancti Hilarionis*, the life of St. Hilarion by St. Jerome.

In those pre-sputum test days just after the Emperor Constantius died (365 A.D.), a Christian by the name of Italicus owned some horses and used to run them against the horses of a certain man who was the duumvir of Gaza and a pagan. Italicus' ponies didn't seem to be in the same class as the pagan's: they lost consistently. Word got around that the pagan was performing certain magical rites which were pepping up his horses and slowing down those of his opponent.

Since this was obviously a problem in practical theology, Italicus consulted St. Hilarion. At first Hilarion wouldn't handle the case. But Italicus overcame his objections and obtained a bowl of water that Hilarion had himself consecrated.

The day of the races arrived. The spectators crowded the stands. The chariots were lined up. Just before the starting signal, Italicus sprinkled his horses with the consecrated water. His horses flew. The pagan horses walked, crept, faltered. The price was long, as one can imagine, and some of the crowd rioted and called for the death of the Christian magician. But others recognized the miracle for what it was worth and abandoned paganism as a consequence.

But let me return to the traveller who was on a site visit to the desert.

*Added in 1983: it no longer does.

Shortly after the crocodile incident, the traveller met up with Paphnutius, the holy man of God, the most famous of the anchorites who lived in the remote desert near Heraclios.

At the end of his life Paphnutius prayed to God and asked God to show him which of the saints he was most like. (An immodest request, to my way of thinking.) An angel then appeared and answered Paphnutius that he was most like a certain man who earned his living by singing in the village. The answer confused and upset Paphnutius. He immediately went to the village and sought out the man and asked him what his saintly credentials were.

The man answered that he hardly thought himself to be saint; that up to recently he had earned a living as a thief and now he had sunk even lower and was making out by singing. Paphnutius figured that there must be some little redeeming feature the man had neglected to tell. The thief thought about it for a while and then recalled an incident. Years before when he was working with a group of thieves, they captured a virgin who was consecrated to God. The rest of the group were all for deflowering her then and there. But he snatched her up from the group, sped her away at night and restored her to her house.

The thief recalled another incident.

Once he found a woman wandering alone in the desert. She was hungry and destitute. The thief asked her what had brought her to that state and she answered him that her husband and her three sons were all in jail for back taxes. The authorities were treating them cruelly and were on her trail to throw her in jail for the same debt. (Lest this story appear too archaic, I remind myself of the stories I keep hearing of how the poor are often ground to the earth by collection agencies that put out every snare to lure them into debt in the first place.) The thief took the woman to his cave where he fed her and then gave her the money necessary to release her family.

When Paphnutius heard these stories he was plenty impressed. He admitted that he had done nothing like these good deeds. Nonetheless, he had been operating in his own sphere and had piled up a reasonable amount of merit in his own field. And then Paphnutius, proselytizing the thief, said to him, "And so, Brother, seeing that you are not occupying the lowest room in God's House, do not neglect your soul." Whereupon the thief flung away his musical instruments and followed Paphnutius into the desert. There he was able to transform his musical skills into spiritual harmony. He spent his days in abstinence and prayer and at the end of three years gave up his spirit amid the angelic host of the saints.

The Paphnutii of the world are all made of tough stuff. You would think that this revelation would have satisfied him. But no. He again prayed to God to show him his like on earth. I shall not go in to what was revealed this time. It is easy to imagine similar stories. Besides, I am anxious to introduce *the* literary Paphnutius, par excellence.

VIII.

THAÏS

The world was created, so one legend goes, from the union of opposites. Since true opposites, cannot, logically speaking, exist simultaneously, both must give way, and in the collapse, new forms are created. This is the dialectic of the legend. In mathematics, plus infinity and minus infinity are, in a sense, opposites. Out of their conflict, symbolized by $\infty - \infty$, comes the theory of limits. The mathematicians are clever — they can turn this struggle into ∞/∞ or into o/o — and out of these forms comes the whole of the differential and integral calculus.

In the less cerebral world, the opposites are only approximate: the cobra is pitted against the mongoose, the man of God against the whore. The last fight is eternal; it symbolizes and preserves the split of human existence into the spirit and the flesh. (Oh, said one of my young colleagues who had been recuperating from an operation, who needs the goddamn body? It's a pain in the ass. If only I could get my mind clear again, I'd be able to get down and do some work.) You will therefore not be surprised to learn that one really can't pursue tangential historical matters for too long a period without coming up against the story of the whore and the man of God. The name of the story is

Thaïs. It's also called Asenath. It's also called Sadie Thompson. It also goes by the name of the State of Rhode Island and Providence Plantations v. Avery, heard in the Supreme Judicial Court, 1833.

I picked up the thread from Paphnuty to Thaïs in this way. My children, of course, had long been alerted to my hobby. One evening, I got a long distance call from my daughter who at that time was studying Classics in Philadelphia. She told me to look up Hrotsvitha of Gandersheim and I would be pleasantly surprised. The telephone connection to Philadelphia was bad and it took several rounds of spellings before I got the name straight. The clue was a bit on the coy side, but hearing is obeying, and I was off on a hunt without too much delay. I found nothing in the *World Book*.

After some digging around, I established several facts. In the first place, Hrotsvitha (around 935-1001) was a nun. She lived in a convent of Gandersheim, which was in Saxony. In the middle of a century known as the nadir of everything, Hrotsvitha studied the classics very hard, particularly Terence, and turned out a number of very remarkable plays in Latin. I established further, that dumb as it may sound to us, there were at least two Hrotsvithas of Gandersheim (popular name, evidently, among the Saxon girls, meaning 'strong voice,' so one can imagine the type), and the authorship of the plays is in dispute. The situation is much more serious than with Shakespeare and Bacon, for what is at stake here is whether the plays were written by Hrotsvitha of Gandersheim or by Hrotsvitha of Gandersheim.

I learned more than this. I learned that there was a Hrotsvitha Club in the United States, founded some-

Map of old Gandersheim. Engraved plate from Johann Georg Leuckfeld, *Antiquitates Gandersheimenses*, Wolfenbüttel, 1709.

Thaïs, American style. (Drawing by Frank C. Papé, published in *Thaïs*, Anatole France. New York: Dodd, Mead & Co., 1926.)

time in the 1920s, and wonder upon wonders, one of the high officials in the club was a woman, retired for some years now, who had been a curator of rare books at my University. The function of the club: to promote the interests of rare book collecting in general, to publish material about Hrotsvitha in specific, and to act as a kind of feminist organization in the post-suffragette period which was a nadir of women's activism.

Still no connection to Pafnuty. I decided to take the bull by the horns, as it were, and to write to Miss Wallace, in Greene, Rhode Island. Miss Wallace very graciously came out of retirement and answered me. She wrote that she was very pleased I was interested in the work of the Hrotsvitha Club, and though its numbers had been sadly depleted by members passing on and by other distractions of life, she still did a little club work from time to time. She sent me by return post a very beautiful privately printed and bound, but uncut, edition in English of the *Collected Plays of Hrotsvitha*. The mystery resolved itself as soon as I was able to cut the pages. On page 83: the play *Paphnutius*. Dramatis Personae: Paphnutius, a hermit. Disciples of Paphnutius. Thaïs, a Courtesan. Lovers of Thaïs. Anthony and Paul, hermits of the Thebaid. An Abbess. I learned further that this was the first of Hrotsvitha's plays to be translated into English (1913) and shortly thereafter it was performed at the Savoy in London with Ellen Terry taking a small part. Ellen Terry was 65 at the time. Just right for the Abbess.

Well, once I was on to Thaïs, it was not hard to connect up with Anatole France's Thaïs, Jules Massenet's opera Thaïs, and thence all the way back to the Fourth century when Thaïs stories abounded in all kinds of variations and with a variety of endings. The whore turned saint is always called Thaïs, or some close variant. The man of God is Paphnutius or Serapion in about equal proportions. In the Massenet opera, Paphnutius is called Athanael (much to Anatole France's disgust), because Massenet's librettist could find only two rhymes in French for Paphnuce.

The standard Thaïs story is the one given in the Vitae Patrum. It has such a nasty ending that Helen Waddell couldn't bear to retranslate it and she left it out of the Desert Fathers. In brief, the story is this (I, too, can hardly bear the ending). Paphnutius, a fanatic, to say the very least, hears about Thaïs and deliberately sets out to change her evil ways. He succeeds, and as part of her repentance, Thaïs burns all the goods and finery she has amassed. Then Paphnuty seals her up in a small cell. After three years, it occurs to Paphnuty to take a look and see what is happening back at the cell. There is Thaïs sitting in her filth, looking toward the east. She is saying over and over again: "Qui plasmasti me, miserere mei; thou who hast created me, take pity on me." Paphnutius releases her. She lives for 15 days and dies.

What to do with the ending? Every writer who tackles the story is confronted with the dilemma. The solutions vary as widely as the personalities of the authors. Oscar Wilde wrote a Thaïs. He lost the manuscript in a taxi. His efforts to reclaim it proved fruitless and he never had the interest to rewrite it. But he used to tell his version of the story to his friends at great length. Wilde's solution was that at the end, Paphnutius loses his faith. He proceeded on the general supposition that whenever A converts B, A's own strength of faith is drained away.

In 1899, with Anatole France's Thaïs on everyone's lips and the succès fou of Jules Massenet's Thaïs dinning in everyone's ears, a tomb was opened by archaeologists in the Graeco-Byzantine necropolis of Antinoë in Lower Egypt. The tomb had a chamber about six feet long and two and a half feet wide. It was built of stones cemented together and was topped by an arch. There was a niche at the head, and on it was an inscription written in red ink. Only part of the inscription survived:

> \+ Ekoimethema
> Karia Thaias
> . . . Thessal . . .

The first two lines can be translated as: *Here lies the blessed Thaïs.* In the crumbling coffin, the remains were surrounded by a number of religious objects not characteristic of other tombs in the same cemetery. There was a sacrament case of palm fibers, an ancient rosary of wood and ivory, a rose of Jericho, the symbol of immortality and resurrection, and, found between the fingers of the skeleton, a cross in the form of an ankh, the symbol of life and rebirth. There were also some palm leaves signifying a martyr's triumph.

The question immediately flew to everyone's mind: was Thaïs a legendary figure or did she really live? If so, was this her grave?

Well, if this were Thaïs, why not look around for Paphnutius? They did. And there are some who think

Monks, Egyptian style

he was found. Not far from where Thaïs lay, a grave was opened. It contained the remains of a monk. He was dressed in a brown robe, a black cloak. He had sandals studded with nails. He had a belt, a staff, bracelets and anklets of iron, and a cross suspended from a collar. The archaeologists looked for an inscription. They found none. But on a potsherd lying close by they found words: Serapion, (son of) Kornosthalos. Not Paphnuty, but Serapion. Believe it or not, as Ripley used to say!

The archaeologist came to the conservative conclusion that he had found no evidence to connect these remains with the legendary figures. On the other hand, he insisted that he had found no evidence to disconnect them. Arguments from silence are pretty weak; nonetheless, the matter remains open.

IX.

CADBURY

It must not be thought that I have spent the past ten years of my life pafnutying. (I should like to be remembered as the man who introduced this verb into the language. I think it has great possibilities, like Oscar Wilde's 'bunburying.' In future editions of Webster, I should like to see: to pafnuty, v.i., to pursue tangential matters with hobby-like zeal.)

Most of the time my fires are banked and burn low. Sporadically, someone applies the bellows. A serious outbreak of pafnutying occurred in the early months of 1970, accompanied by circumstances that are sufficiently mysterious to warrant a separate investigation on their own merits. To explain the circumstances properly I must go back to the spring of 1969 when I received a letter addressed to me at Brown University, Providence, Rhode Island, New York, U.S.A. The letter was postmarked Hobart, Tasmania, and this, I supposed, was sufficient to explain the interesting location of Providence. For my part, I wasn't quite sure whether Tasmania was in Africa or Australia. I thought the latter, but I wasn't going to put money on it.

The letter was from a man whom I shall call Dugald Campbell. I knew Campbell professionally, but I had never met him. His letter invited me to come to Tasmania as the guest of the Australian Mathematical Society and give a series of lectures at their Summer Conference in January, 1970. Ah yes, down under, winter is summer. This was too nice an opportunity to turn down so I prepared my lectures and packed my bags.

One day I found myself landing at the airport of that remote island just south of the eastern Australian mainland. Van Dieman's Land it used to be called, the Ultima Thule of the Southern Hemisphere; its valleys,

I thought, would be populated by the Tasmanian Devil and its shores lapped by the melting ice cubes of the Antarctic. As I walked down the steps of the plane looking toward the gate for someone who might be looking for me, I fully expected to be greeted as St. Anthony was greeted by Paul the Hermit in the mountainous deserts of Egypt: "Tell me, I pray thee, how fares the human race: if new roofs be risen in the ancient cities; whose empire is it that now sways the world; and if any still survive, snared in the error of the demons."

But none of this; instead, families meeting weary travellers, who were returning home from visiting relatives on the mainland and carrying with them the first fruits of Melbourne department store sales.

My lectures, I thought, went well, and I had a chance to see quite a bit of the island. Tasmania is a beautiful place, rather mountainous but with many green portions; slightly reminiscent, in my eyes, of Vermont. Hobart, the capital city and the location of the University of Tasmania, is beautifully situated on the Derwent Estuary. The hills come right down to the water, and Mt. Wellington lords it in the background. Practically every house is an estate agent's dream and has a panoramic view of the Derwent opening into the Bay of D'Entrecasteaux. The waters are dotted with white sails.

The day before I left Tasmania I was the guest of honor at a luncheon at the Royal Tasmanian Yacht Club across the road from the University. The party consisted of the Vice-Chancellor of the University, Sir Arthur Wrentham and Lady Wrentham, Professor and Mrs. Campbell, Professor Vahler of the University of

Manchester, who like myself was a visiting lecturer, Professor William Collins from Adelaide, and Mr. and Mrs. Smythe-Jones. Mr. Smythe-Jones was the resident manager of the Cadbury Chocolate Factory in Hobart and as the conversation deepened, it emerged that Cadbury's had contributed something toward defraying the costs of the Conference. I supposed that Nougat and Heavenly Creme would shortly be computerized. Mrs. Smythe-Jones was the only native Tasmanian at the table. She was Smythe-Jones' second wife.

I cannot remember the exact flow of the conversation, but as one part is critical to my story, I shall have to sketch it as best I can. I asked Professor Campbell whether the people in Tasmania knew much about Rhode Island. He answered that they did indeed know quite a bit about Rhode Island, New York because Tasmanians were mad for sailing. They followed the America's Cup Race run off Newport with great interest, particularly when it involved an Australian contender. They also followed the Newport to Bermuda Race because the Australians have a similar race — Sydney to Hobart, about 600 miles. It turned out that this year's winner was now in dry dock and he pointed out the window to a beautiful boat having a hull job done.

Sir Arthur Wrentham, it appeared, was not an educator but a career administrator with many years in the colonial service. He had played an important role in the transition of Nigeria and Uganda to self-rule and had won the Order of the British Empire for it. His present post was an amusing relaxation between jobs. Lady Wrentham concurred. Sir Arthur went on to say that the British had ruled the whole of India with 1,000 soldiers and much bluff. In those days everyone was civilized and knew it was bluff. But the game had changed.

Professor Varler said "Hear, hear," and that he, as a native of pre-World War I Germany, but with a mathematical view of history, could only wish for a return to the stability of the Pax Britannica. And so it went for a while until the sweet was brought in. I recall it was a delicious fruit salad slopped over with the ubiquitous thin custard sauce. When the empires that now hold sway over the world shall all have fallen into dust, some future anthropologist will be able to trace the extent of the British Empire by examining what people of nations now unborn have a predilection for custard sauce. Custard sauce accomplished as much for British Imperialism as stone walls did for ancient Rome.

The sight of this yellow substance strangely reminded me of Scotland. And Scotland reminded me, in the proximity of Professor and Mrs. Dugald Campbell both of whom were Scots, of Professor John Begg at St. Ida's. I asked Campbell whether by any chance he knew Begg. He did not. But Professor Collins, who had been silent all the while, spoke up. "Know him?" he answered, "I did my thesis under Begg." And then the knife was plunged in and twisted slowly. "Incidentally," he said,' "when you wrote your lovely book on interpolation and approximation why on earth did you spell Tschebyscheff's name in such an outlandish fashion? I always spell it Chebichef!"

Oh brother, here we go again, and in Tasmania, 12,000 miles from no place, where I had met up with gentle American students escaping from the bomb and expecting to be fed by birds like St. Paul the Hermit.

But I dodged the issue; there were ladies present and one should not talk shop. Besides, there was a man from the chocolate plant present, and if the mathematicians couldn't agree on a simple spelling, how wouldn't they bugger up his cremes and kisses once he had opened up his kettles to their computers.

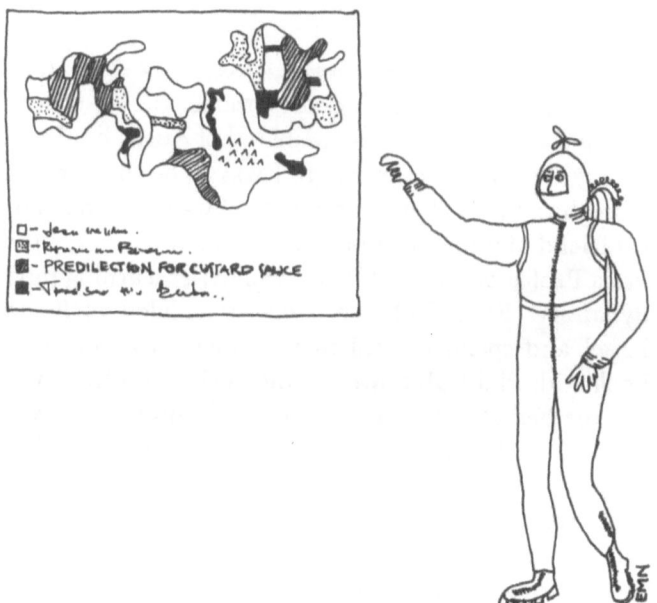

Anthropologist of the future traces the extent of the British Empire by examining what peoples have a predilection for custard sauce.

X.

YETI

My plane became airborne and flew north. Soon it was over the Bass Strait separating Tasmania from the mainland of Australia. I was on my way to Teheran where I was to be the guest of Abolghassem Zaghari's family. Abol was a mathematical friend of mine; together we had sat at the feet of the great Professor Richard Ritter von Mises learning fluid dynamics. From Teheran I was scheduled to fly to Israel and spend several months there working on a book with Phil Rabinowitz of the Weizmann Institute. We mathematicians, possibly weak in numbers, have a Masonic fellowship which extends across the globe. Joined by a system of research papers and its apparatus of mutual references and cross references, occasional collaborative work, and personal friendship, we can travel from continent to continent with reasonable expectations of finding a flop, a meal, and an invitation to present a colloquium talk.

What had I gained from my trip to the antipodes? Much, I thought in terms of personal friendships made. But not the least of it was the fact that I had finally seen for myself that the Australians walk around upside down. Now you cannot tell this just by looking at them. They appear to go about their business pretty

The moon as viewed by Tasmanians.

much as we do in Providence, R.I. Their feet seem to be on the ground, their heads seem to point toward the heavens. How, then, can you tell? The easiest way is by looking up at the moon. Remember what the moon looked like back home. Remember the markings. Now try to find the man in the moon. You can't. He is upside down. You are led irretrievably to one of two conclusions. Either someone turned the moon upside down while you were on the plane or you are standing on your head. Take your pick. When, at the end of days, Uncle Sam and the Russian Bear join hands in amity and sweet reasonableness to erect a Vodka and Coke sign on the moon, in all likelihood it will be erected so as to be readable from the Northern Hemisphere. The people living in Australia, New Zealand and Patagonia, all of them standing on their heads, will not be able to read it. The sign will be upside down. They will complain. The Battle of the Hemispheres will have begun.

What is belief? What is evidence? What is witness? What is testimony? What is truth? The reader will allow me, I hope, while my plane is speeding westward from Adelaide to Perth on an otherwise uneventful trip, to introduce a few side issues, a few irrelevancies to a main story which is itself the King of Irrelevancies. The great Irish historian Lecky whose statue I once saw in the courtyard of Trinity College in Dublin, wrote a book called *A History of the Rise and Influence of the Spirit of Rationalism in Europe*. This title will be rather more comprehensible if I point out that the first part of the book is subtitled, "On the Declining Sense of the Miraculous." Lecky's book is based upon the observation that until very recently the peoples of the earth, what-

W. E. H. Lecky

ever their degree of primitiveness or sophistication, whatever their religion, whatever their politics, whatever their station in life or their location on the face of the earth, were all believers in the supernatural. They believed in ghosts, devils, in witches, in miracles, in relics, in supernatural events. These were not rare occurrences; they were everyday events and all had seen such manifestations or had spoken to those who had. Gradually, over the centuries, this belief turned to disbelief. Lecky, writing from the high point of a strong mid-Victorian belief in progress and scientific and rational thought, set himself the task of describing the slow transformation from supernatural to rational thought and of answering why the transformation occurred.

Lecky wrote: "When so complete a change takes place in public opinion, it may be ascribed to one or other of two causes. It may be the result of a controversy which has conclusively settled the question, establishing to the satisfaction of all parties a clear preponderance of an argument or fact in favor of one opinion . . . It is possible, also, for as complete a change to be effected by what is called the spirit of the age. The general intellectual tendencies pervading the literature of a century profoundly modify the character of the public mind. They form a new tone and habit of thought. They alter the measure of probability. They create new attractions and new antipathies, and they eventually cause as absolute a rejection of certain old opinions as could be produced by the most cogent and definite arguments."

The decline in the belief in the miraculous, says Lecky, is due to this second cause, the Zeitgeist. And I must say, he makes a good case for his position. His study was written before the industrial and scientific revolutions intensified, before the Darwinians had "vanquished" the clergy; before even, as has happened in our generation, cracks had begun to appear in the plaster of rationalism. To explain a pervasive feeling of mankind in terms of the spirit of the age, in terms of a predisposition to believe or accept one point of view as opposed to another, strikes me, I admit, as somewhat tautologous, but it may be the best we can do. Science, says Newton, cannot explain; it can merely describe one mechanism in terms of another. History can do no better.

One September day not too long ago, visitors to the Theodore F. Greene Airport in Providence, Rhode Island would have seen deplaning a young man of oriental cast wearing a flowing coat and skirt and burdened down with two heavy grips. The young man was Sikkimese and had just arrived from that remote land.

Sikkim is a small country in the Himalayas lying between India and Tibet. It is the closest that the world comes to Shangri-La. What one of its sons was doing in Providence is a story that starts when the world was created, but for brevity, let us say only that the Queen* of Sikkim is an American. Through her American connections, a scholarship was set up at Brown in support of Sikkimese students. The young man coming to our house was the first such scholarship student, and we were putting him up, as do many faculty with foreign students, until the dorm opened.

I shall call him Sam. After a few days, the travel weariness wore off and some of the culture shock. My wife and I saw immediately that Sam was a charmer. He spoke English very well indeed. He had a slight Indian accent and a tendency to omit the articles. He had unusual warmth and openness of personality. He made himself at home immediately and became almost one of the family.

Though it was a little bit difficult to tell, Sam seemed satisfied with our food. The main objection seemed to

*Added in 1983: Ex-Queen

be that it was too soft and not peppery enough. We brought out our bottle of Tabasco sauce which he applied liberally. We promised to buy a bottle of hot peppers for his use.

Sam visited us from time to time during the semester and would stay for supper with the family. One such evening the conversation at the table was memorable. Sam had brought along an illustrated paperback on karate and before supper went out to the backyard to practice for a while. He said he was a beginner and didn't know how much progress he could make by himself. My son Ernie asked him what the great masters of karate could do, and he answered that he had heard that they could split a thick plank or a brick with their bare hand. Someone suggested that the karate master had in some way secretly prepared the plank beforehand so that it would split easily. Sam answered that it was not done by conjuring but required a perfect, almost trance-like state of mind on the part of the karate master.

My son Joey asked Sam whether there were any magicians in Sikkim. "There are Shamans," he answered, "and they do wonderful things." "like what?" "They can cut a chestnut leaf with a feather." "You've seen it done?" asked Frank. "I have."

Ernie, who likes to unify all human experience, said that in the days when the Aztecs in Mexico offered human sacrifices, the victim was given a choice: he selected either to be sacrificed straightaway or to fight his way to freedom. If he selected the latter option, he was armed with a plume and had to face twelve men who were armed with weapons of war. Ernie added that no cases of escapees were recorded. Sam said that the Aztecs must have been a very cruel people. The Sikkimese are a very gentle and friendly people. There are no jails in Sikkim.

"What other kinds of things do the magicians do?" asked Joey. "I'll tell you another thing. I don't think it is magic, but it is a miracle, and I have seen it. The Sikkimese are Buddhists," said Sam, "and we put out cups of water as an offering. People do this in their houses and also in the temples.

"Once a year the King goes to a certain temple to perform the miracle of the cup that is located there. The priests in the temple lead him to a special cabinet, very beautiful. The cabinet has one door. The door is sealed with the royal seal. The King breaks open the seal and opens the door. Inside the cabinet there is a cup. It is full of water. The King pours out the water as an offering. He returns the empty cup to the cabinet. He reseals the door with the royal seal. When he comes back next year, the cup will be full. It is a miracle."

Frank took issue. There were many ways of explaining this naturalistically. It might be condensation. Or it might be just plain trickery. "Our great magician Houdini," he said, "devoted a lifetime to debunking the supernatural. The locked box trick is nothing, you can get the equipment for it in any joke shop."

"No," said Sam, "it is not a trick. It is a miracle. We have smart people too who examined the situation. They are unable to explain."

Ernie asked Sam whether the magicians can walk on hot coals. "They can," Sam said, "I have seen it. First they put themselves into a trance by beating on drums for many hours. Then they walk on coals. They are in a state of mind when they do it. I will show you something," he said to Ernie. The bottle of crushed hot peppers was on the table. Sam sprinkled some of the pepper onto his palm. "You see, these parts here are the dried flesh of the pepper. They are hot, but not too hot. Now this little thing is the dried seed. The seed is very hot." He gave a single seed to Ernie and told him to bite into it. Ernie did so. In about twenty seconds, the heat from the seed spread over the wall of his mouth and Ernie got up from the table howling with pain and ran into the kitchen to rinse his mouth.

"That was from just one seed," said Sam. "Now watch me. Sam took a slice of salami from the cold cut tray, and dusted it over liberally with hot peppers so that the pepper seeds lay about an eighth of an inch thick on the meat. Then he ate it with relish. "Not too bad," he said, "but the peppers in Sikkim are nice and fresh. Much hotter."

Frank said that what Sam had done was nothing, as Sam had been trained from childhood to eat hot things. I added that I knew men who could drink me under the table. "Maybe it is also preparation of the mind," answered Sam. "In the West, when a man gets very old, they send for doctor to help his body to live longer. In my country, they send in priest to help him prepare mind to die."

"Who trains the magicians?" asked Joey. "They have long training. They are stolen by the Yeti when they are children and taken into the woods and hills." "What are the Yeti?" asked Ernie. "The Yeti are the Abominable Snowmen. They live in the mountains just at the snow line." "Have you seen one?" Ernie pumped. "No, but I have been in the small villages where the men in the village have seen signs of the Yeti.

"The Yeti train the magicians. Then they come back to the villages after some years. But the captured children must be protected from the female Yeti who would eat them if she could. The children are kept in cages."

"What do the magicians say about the Yeti?" "They don't say." "Well, what evidence of the Yeti do the villagers find?" "Footprints, yaks mutilated and partially eaten."

"Couldn't that have been the action of a wild animal, like a wolf or a bear?" "No," said Sam, "I have heard of yaks found strung up high in trees. An animal cannot tie up a yak, and string him. Men in village have no reason to do so. This was done by Yeti."

This is an accurate transcription of a conversation held in Providence, R. I., in the Year of Unbelief 1972.

What is reason? What is belief? What in fact do we see? How do we interpret what we see?

XI.

LAMA TED

Shortly after Sam graduated from college, I ran into him on the streets of Providence. I asked him what he was doing and he told me that he was housepainting to earn money. I asked him whether he had any plans for the future and he answered that he had many irons in the fire. He was giving talks on Sikkimese Culture to Rhode Island school children, he was translating Nineteenth century Lamaistic philosophy into English, he was helping to build a Lamaist Temple, he was getting married, he was thinking of setting up an import business, he was trying to sell mandalas, he was also seriously thinking of going into retreat for a nine month period.

"Where would you go into retreat?"

"In Providence."

"That should not be difficult. Our detractors say that the whole City is in retreat. They are just jealous. The City is in what one might call contemplative but creative relaxation."

"Yes. It is very good place for thinking about Universe."

I asked Sam what a mandala was — a fruit? — an instrument? It turned out that a mandala is a mystic religious drawing that is used as an aid to contemplation or for purposes of magic. They are often drawn in sand. He promised to bring me one in a few days, and sure enough, he came round to the house a few days later bearing gifts of honey and granola and numerous samples of mandalas.

The mandalas were about ten inches square. An outer circle encloses a square whose interior is divided with mathematical precision into labyrinthine chambers in which various gods, scenes, designs, patterns have been carefully inserted. Sam's mandalas were

Mandala.

hand painted on cloth in beautifully clear color. The whole production, from the geometry and the iconography to the painting process, takes much knowledge and many months to execute.

To some extent the mandalas reminded me of the "ojos" or "eyes" of the Pueblo Indians, religious symbols that one finds in New Mexican art shops, but the mandalas exhibit a much finer structure and come with a correspondingly more elaborate explanation of their components. Displaying the mandala of the tutelary wrathful deity of palchen dus pa, and preceding from its central core to the outer circumference, Sam pointed out the sphere of the gods, the wall, the ramparts, the basement, the four passages, the areas of darkness and

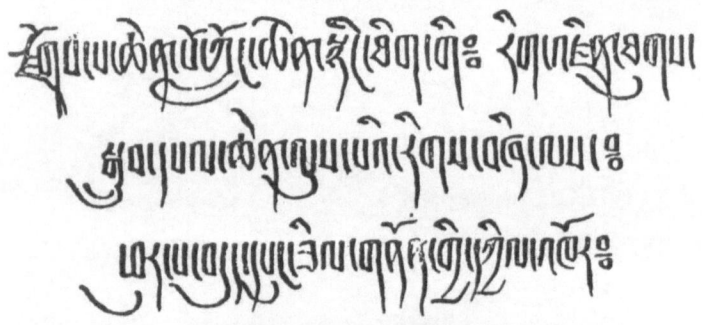

RDZOGS PA CHEN PO
KLONG CHEN SNYING THIG GI
RIG 'ZIN THUGS SGRUB DPAL CHEN 'DUS PA'I
RIGS BSHI LAS
PHUR BA BDUD DPUNG ZIL GNON GYI
DKYIL 'KHOR.

Brief explanation of mandala.

obscurity, and so on until he came to the belt of fire that circumscribes the whole figure. The inner portion also reminded me of the board of a pachisi game and I wondered whether this children's game had, in previous centuries, served as a miniaturized and stylized version of the game of life itself. I knew that in Vedic times, the kings ritually cast dice — often loaded dice.

This particular mandala, Sam told me had been made in Providence by a lama. I expressed surprise to learn that there were any lamas in Providence, but Sam turned away my surprise by saying quite coolly, yes, he himself had attracted two lamas, and that due to the oppression and dispersion of the Dalai Lama and the lesser lamas by the Communists in Tibet, Providence had become one of the leading centers of Lama-istic Buddhism in the United States.

The man whom I shall call Lama Ted came directly to Providence from Sikkim. One sees him occasionally on the street wearing saffron robes and a triangular saffron cap. He works in the kitchen of a sandwich shop popular with the art students. I asked Sam how Lama Ted had managed the culture shock of moving from Sikkim. Sam answered that it was quite easy for him as the exterior world, whether in Asia or in North America, was of secondary interest to him. What distressed him somewhat was the reverse culture shock he sometimes induced in unwary Americans. When he first arrived, he was put up in a room in an area where many art students live. One morning he went to the window and saw a young woman sitting in the back yard chipping away on a large log. She appeared to be creating a fierce profile of totemic aspect. Lama Ted became interested in what she was doing and went down to the backyard and quietly watched over her shoulders. The woman chipped away at the profile unaware, but finally she turned, and when she caught sight of her exotic onlooker in full regalia, she screamed and ran into the house. Lama Ted was left as the tutelary guardian of the wooden ikon.

But the great story of his first days in Providence, said Sam, was how Lama Ted prevented a hurricane from devastating the East Coast of the United States — from Eastport to Sandy Hook. It came about in this way. Arriving safely in Providence after a long and tiring flight, Lama Ted spent several weeks resting and settling himself in. At the end of this time, which must have been around the beginning of October, he told Sam that he wanted to commemorate his safe trip by sacrificing to the Spirits of the Waters. For this purpose he would need convenient access to the ocean. Sam took him down to India Point Park which borders on the northern tip of Narragansett Bay. Technically, this is ocean with a tide of several feet whose rise and fall is tabulated in the Providence Almanac. There are sea gulls and all that, but Lama Ted decided that the ocean at Providence was not sufficiently extensive and open and that in any case he would need a place such as a bridge where he would be above an expanse of water. Sam told him about the Mount Hope Bridge about twenty miles south of the city. Lama Ted thought this place would be acceptable and told Sam he would go by bus the following week on a day to be determined ritually. His plan was to buy a round trip ticket to Newport, get off at the bridge, perform the ceremony and return.

Mount Hope Bridge.

Three or four days before the set day, the radio began reporting that a storm of hurricane proportions was making its way up the Atlantic Coast from the Carribean. The storm was being watched carefully and given a name — let's call it Felicia. Its cyclonic wind velocity was determined to be very high indeed and its movement northward was carefully plotted and projected forward. Felicia was expected to pass over Long Island and Southern New England on precisely the afternoon that Lama Ted had selected for his sacrifice.

The storm warnings now extended from Cape Hatteras to Newfoundland. The owners of small vessels tied up in marinas were advised to secure their ships and householders along the south eastern coast of Long Island were alerted for possible evacuation. The morning brought a steady downpour of heavy warm rain and the winds rose steadily in strength. Travellers' advisories were pronounced. By eleven o'clock in the morning, businesses and institutions in the Rhode Island area allowed their people to go home. Schools and scheduled events were cancelled. Radio Cassandras reminded us to lay in a supply of candles and canned goods and to fill our bathtubs with fresh water. The winds approached 45 with occasional gusting to 65. The barometer fell below twenty-nine inches. The streets were strewn with large branches snapped from trees. Flooding in cellars was widely reported. The keepers of the hurricane gates just below downtown Providence lowered this barrier against a possible tidal wave up the bay such as had inundated the city in 1938 and once again in 1954.

In the middle of this storm and meteorological apprehension, carrying an umbrella which the winds soon made useless, Lama Ted found his way to the central bus station in Providence to catch the 1:00 P.M. bus to Newport. Sam says that at that time the Lama had an English vocabulary of perhaps a dozen words and that he had drilled him in three particular words: "Providence," "Newport," and "Mt. Hope Bridge." The bus was strangely on time, Lama Ted bought his round trip ticket and told an incredulous driver, "Mt. Hope Bridge."

Acquidneck Island, formerly called Rhode Island — this is the original Rhode Island while the mainland portion of the state was known as Providence Plantations — is a triangular island about fifteen miles in length, on which there are three cities: Portsmouth, Middletown, and Newport. The island is reached by three bridges, the great Newport suspension bridge from the west and the Mount Hope and Tiverton Bridges from the north and east. The Mount Hope Bridge joining Bristol on the mainland and Portsmouth on Acquidneck is also a suspension bridge, the second largest in New England, more than a mile in total length with a central segment better than a hundred feet above the passageway joining Mount Hope and Narragansett Bays. The bridge has a toll barrier on the Bristol side and is not customarily open to pedestrians.

The bus made its way along the highway in lakes of water. Believing that Lama Ted could not conceivably have any business at the bridge itself, the driver let him off at the college just short of the bridge. Lama Ted found his way through the warm driving rain to the bridge entrance, past the toll booth and up the span to the highest position. There he performed the ceremony of sacrifice to the Spirits of the Waters. As a part of the ceremony, he cast off onto the waters long paper streamers and flags on which prayers and verses were written in Sikkimese and Tibetan characters.

A car proceeding northward to Bristol spotted him and reported at the toll gate that a nut dressed up in a costume was at the middle of the bridge doing some very strange things. Perhaps it was a case of suicide. The toll gate called up the Bristol police and within a few minutes a patrol car with a siren and a flashing red dome light picked up Lama Ted and brought him down to the Portsmouth side. Here was a strange bird indeed, but obviously no nut. Lama Ted now used his one relevant word over and over: "Providence, Providence." The Bristol police, as a courtesy to strangers — I doubt if they realized he had travelled from the foothills of Kanchenjunga, the second highest mountain in the Himalayas, to Mt. Hope, the second longest bridge in New England — were all set to drive him back to Providence in the storm and indicated as much by grunts and signs. The Lama caught their meaning and showed them the return portion of his bus ticket. The police then kept him in the Portsmouth station and flagged down the next bus back to Providence.

In·the meanwhile, the prayer strips cast from the bridge were buffeted by the winds of gale strength but found their way down to the waters and lay upon the face of Mt. Hope Bay. From there the message of the verses passed through Narragansett Bay to Rhode Island Sound, thence to the open Atlantic, possibly to the Antilles to the south or to Newfoundland to the north where the Spirits of the Waters may then have been residing. The winds veered and receded. The mechanism of the hurricane ground strangely to a halt. By the time Lama Ted was back in Providence, the rain had stopped and the sun was shining.

The next morning the Providence Journal reported that the hurricane had abruptly turned eastward, sooner than predicted, and had blown out to sea with only minor damage. The reason for this perturbation was not given, nor was the reason perceived as clearly as Sam had done.

What is reason? What is belief?

Years ago, William James gave a public lecture and said, "Many of us in this hall believe in democracy, in liberal Christianity, and in the existence of the atom, all for reasons that are not worthy of the name."

Should he have listed the Yeti and the Spirit of the Waters?

XII.

NADRA

I did not get to Teheran. The plane landed in Bombay where there was an unexplained four hour delay. It then took off and made an unscheduled landing in Delhi; there were four more hours of delay. At this point we were informed that Teheran was blanketed by the worst snowfall in the history of the BOAC. We would fly to Ankara, refuel, take off and land at Tel Aviv, the next scheduled stop. I was aching with tiredness. My body was confused and fatigued by the constant pushing back of the clock. I decided to get off at Tel Aviv and not wait around for an unknown number of hours and fly back to Teheran. A word to an efficient hostess accomplished this and, miracle of miracles, my baggage was pulled out and given to me personally. After forty or so continuous hours of travel from Perth, with my stomach and bowels standing on their heads, my ears suffering from decompression, my mind outraged additionally by culture shock, I found myself sitting on a bench at a bus stop unable to decide what to do next and not knowing how to do it. I was four days too early. A young boy was next to me eating a bag of sunflower seeds. He was an island in a sea of spat-out shells. This was my introduction to the Holy Land. I had never been there before.

Locally, it was two in the afternoon. I allowed many busses to pass by while I dithered. I decided that I would go up to Jerusalem and spend the four days there. Then I would come down and go to work. I would check into some place, rest up, and then go to The Wall.

The decision itself gave me the strength to act and I decided to hop on the next bus to Jerusalem. A priest in a black gown and a black chef-like hat walked slowly across the road from the airport. He sat down next to me. He began a conversation, but I didn't understand. He fished around in one of his bundles and extracted a small calling card. I read it. It said,

Benedictos III

Exilarch of the Enochian Dispersion

I figured him for a member of some Eastern Rite, but the card was in English, so I tried English again. It didn't work. He tried another language, perhaps it was Greek or Turkish. I tried French. Où allez-vous? Bonjour. Things like that. French didn't work. He tried another language, perhaps Roumanian or some kind of Slavic. I tried pidgin German. He answered me in pidgin German. Click, click, and we were communicating. He told me he wanted to go to Jerusalem and that he needed to go relieve himself. I said I was going to Jerusalem also; we could get on the same bus together. I told him that I thought the nearest men's room was back at the Airport. He got up to go, but he left his eight bundles with me and told me to watch them. I missed one bus before he came back.

Another twenty minutes to wait. Benedictos III told me that he had just come from Romania. It was his first trip to the Terra Sancta and when we got to Jerusalem would I please take him to the Enochian Patriarchate. I said why not, and I thought to myself, I'm not doing anything in particular. As far as sightseeing is concerned any place is as good as any other.

For my part, I told him that I was in the country to work for several months on mathematics. Then I took the plunge. Incidentally, I said, I happened to be interested in St. Paphnutius. Did he know anything about him? Did Paphnutius play any kind of a role in the

Benedictos III, Exilarch of the Enochian Dispersion.

Enochian calendar? My questions were answered with silence. I guess my German must have been bad.

After a bus trip of an hour, one of the fabulous trips in the world, we pulled into Jerusalem. I figured the way to handle the situation was to blow Benedictus III to a cab ride, hoping of course that the cabbie could speak a little English, French, German, Roumanian, Greek or Turkish, and that he would know what and where the Enochian Patriarchate was. I would then dump my charge and ask the cabbie to drive me to some hotel.

It did not come to this. We schlepped our weary way to the street, I with three suitcases and he with eight bundles, and I looked for a cab. We were standing at the curb and a Honda vroomed toward us. A young man was driving; on the rear seat there was a young woman in dungarees and belt with an ankh buckle. She wore a black halter and had her arms wrapped around the young man's waist. The traffic light stopped the Honda. The young man looked at me and recognized me. "Professor Davis? Right! Just great!" the two jumped off. "Don't you recognize me? I was in your Applied Math class four years ago. My name is Jack . . . You gave me a B plus."

There is nothing as discombobulating as to be recognized by a stranger in a strange place. I wouldn't say Jack looked unfamiliar but I couldn't honestly say that I remembered him. Nonetheless, friendship offered is friendship gained. We shook hands. He introduced me to his wraparound. "This is Nadra Ibrahim, Professor, my assistant. This is Professor Davis. Right! Great! Let's help the Professor, right? Great! Can we take you somewhere?"

The Old City of Jerusalem.

I introduced him to Benedictos III and told him that we were about to catch a cab and go to the Enochian Patriarchate. "Right! Good idea! I'll take you there!"

Jack turned the Honda over to his assistant. He gave her a pat on the bottom and she *vroomed* off. He walked around the corner and came back in a cab. A word to the driver and the cab flew. It pulled up, ten minutes later, at a small house, Turkish style, surrounded by a high wall. Over a bell pull, a tile with words on it in a font I could not recognize. Enochian alphabet I supposed. An acolyte came out and kissed the hem of the Exilarch's skirt. He grabbed the eight bundles and they disappeared behind the gate. Benedictos III was home. I wasn't.

"Now," said Jack, rubbing his hands, "I'm going to take care of you." "I need a hotel," I told him, "I've got to go to sleep." "Right! I've got a great place in the Old City. Don't stay in the New City, Professor. Be exotic; live a little. All you'll ever meet in the New City are your neighbors from Providence."

We got back into the waiting cab. I was too tired to argue. I only had an imperfect idea of the New and Old Cities anyway.

En route, Jack pumped me. What did I want to see or do in Jerusalem? The Big J, he called it. Just tell him, he would see that I saw them. Any place at all. Down to the Big T. Down to Jericho. Over to Samaria. Anyplace.

I told him, I just wanted to look around a bit. I wanted to go to The Wall. Maybe to some kind of a library, so that I could track down Pafnuty a bit more. "What's that?" he asked me. In three sentences, I told him all I had accomplished in that field beginning with Tschebyscheff and ending with the Egyptian Desert. "I'm off Math forever, Professor, but Pafnuty is great. I know a man at the University who has written a whole book on Egyptian tenses. I'll get the Pafnuty info for you. Relax, man, you're in my hands."

The driver pulled up at an ancient gate. I paid him off. Jack clapped his hands a few times and whistled. An Arab boy came up and Jack said in Hebrew, "Take the Professor's bags. I'll give you a lira."

To me he said, "This is the Jaffa Gate, we're going into the Old City." We threaded through narrow ways, down and up and down again. Finally, we stopped at a small entry. A sign read "George IV Hostelry". There was a strong smell of urine. I must have looked appalled. The boy said in pidgin, "Very good place. Many English stay here."

Just then, a *vroom* from a motorcycle. Nadra Ibrahim appeared with the Honda. How she managed over the steps where donkeys had difficulties was the least of the mysteries.

All four of us went into the front room. A man in a fez looked up from a desk. He smiled and spoke to Jack. "Ah, Jack! I am glad to see you. I have prepared the best room in the house for the Professor. He will sleep in the Princess Room." I assumed we were being addressed by K. Denisoglu, Prop. This name was displayed on the desk.

All five of us marched up to the Princess Room. What can I say? It had a bed, and a washstand-laver combination; on the wall I saw a 1965 calendar advertising dried figs from Smyrna. There was a radio. The boy turned it on. A blast of rock music came forth from a station in Jordan. Jack and Nadra did the Boogaloo. Prop. K. Denisoglu sat down on the bed and began a long story in broken English, but far better than my Turkish. The Hotel, he told me had been in his family for five generations. It had seen Great Days. Times were now bad and he had to keep skin and bones together with sidelines. In the 1890's when the Princess Royal made a pilgrimage to the Holy Land, her party engaged the whole hotel (four rooms). She stayed in this room and in this bed. After the Great War, when General Allenby marched through the Jaffa Gate in triumph (he, Denisoglu, was just a little boy), his Grandfather bought a new mattress for the Princess Room.

I had had it. Dear friends and children. I have had it. I have just got off the plane from Perth, Australia. I have had innumerable delays. I have avoided the worst Persian snowstorm since the accession of the Pahlevi Dynasty. I have played boy scout to an aged Exilarch. I have checked into a hotel with a far richer history than the Jerusalem Hilton. Let me turn in. Please go, go at once. And do not stand on the order of your going.

"Right! Great!"

XIII.

HEROD

X hours later, where X is an unknown number between 10 and 34, I woke up. I was disoriented. It was dark. The smells and sounds were strange. I looked at my wrist watch I'd taken off and put on the floor. It was no good to me; it was telling Perth time; besides, it had stopped. I found a switch and turned it on. I made my way down a dark hall and found the Princess Royal plumbing. I came back and took a cold water shave. I was shivering. I dressed, putting on two undershirts. I sat in bed in my clothes under the blankets. I tried to go back to sleep, but it wouldn't work. I got up and studied a map of the Old City that I had picked up at the airport. The problem was I didn't know where the hotel was located. The sky was beginning to lighten. I could hear someone downstairs moving about. I could smell coffee and something frying. I put on my coat and ran out of the George IV.

I made my way through the dark terraced streets. I hoped to find some kind of a main street so that I could place myself on the map. Here and there shopkeepers were beginning to take down iron screens from the shop windows. Here and there Arab men in gown and kaffiyeh, and Arab women with their faces covered by yashmaks, went about their business. Donkeys laden with goods were making deliveries. A goat stuck its head from out of an alleyway. I wandered around for a while lost in a labyrinth of courts and pathways. Then I lucked out and found a trilingual sign that said: *To the Western Wall.*

I followed a narrow street. I began to see Chasidim wearing fur streimels, obviously headed for The Wall. I followed them. Two Israeli soldiers with Uzi guns formed a checkpoint. The street opened out to a large plaza below illuminated by the cold yellow sun of the February morning. The Wall stood before me, several hundred feet from the checkpoint.

Whoever has not seen The Wall has not seen Israel; whoever does not understand The Wall does not understand Israel. The Wall: the Wailing Wall, the Western Wall, Ha Kotel Ha Ma'aravi, the portion of the Temple of Herod which still, miraculously shall we say, stands after the destruction of the Temple in the year 70 by the Roman legions. Look at the stones. They are high by today's building standards. Most have a gnarled and eroded surface. On some of the stones the decorative margins of the original dressing is still visible, and one might project back two thousand years and realize what a magnificent structure the original must have been. After the 1967 War, when a plaza was cleared, a considerable length of The Wall became accessible. To the extreme right, there are deep archaeological excavations. To the left there is a spot where the ground has been cut through, displaying something like seventeen layers of building stone below ground level.

The Wall.

What do you do at The Wall? Well, you pray, of course. Or you may simply stand there and think your private thoughts. You may walk about and look at and listen to those who are praying. What prayers do you say at The Wall? If it is morning, say the Morning Prayers. If it is Evening, say the Evening Prayers. You may say your private prayers. You may pray for your family or for someone's health. You may pray for mankind. You may curse mankind. Many people write out requests on little slips of paper and stuff them into the interstices between the stones. From a distance, I saw a black priest in a Roman collar go up to The Wall and take out a little book. I do not know how he prayed.

Does anyone cry at The Wall? Some do. I have heard them. Why should they not? Has no one suffered? Is this why The Wall is called The Wailing Wall? Some people think so. But I have heard a story: it is called the Wailing Wall not because the people cry but because The Wall itself cries. The Wall is a Ruin and cries because it is bereft.

Does the miraculous occur at The Wall? One hears stories, but your interpretation must be according to your predisposition of thought. There are no piles of discarded crutches at The Wall.

Is the Wall a Holy Place? Does the Divine Presence of God hover over it with greater persistence than at other places?

The Sages have said that wherever a single man studies the Torah, the Divine Presence dwells with him.

The Wall is a Holy Place within the definitions of international politics. But in the larger sense walls are not holy.

Why then do Jews assemble at The Wall in multitudes? Why do they insist they must retain The Wall within their governance? It is a symbol for them.

What is a symbol? A symbol is a carrier of meaning and experience. It is an abstraction of the ideal from the real. It epitomizes the conflict between man's finite nature and his dreams of infinity.

And so, in the Year of Unbelief 1970, one hundred years after Lecky wrote his essay on the decline of the miraculous, I, an unbeliever, and a novice at prayer, approached The Wall. A Chassid saw me coming and handed me a yarmulka to cover my head. In the privacy of communion with the stones of Herod, and in the openness of a public plaza, I cried and prayed to God that he confirm me and strengthen me in my Unbelief. And I pushed a little note into a crack. In the weeks that followed, the winds might blow it out, or the swallows might come down in the rain and pluck it out. The beadle might then sweep it away. No matter; it is the process that counts.

XIV.

BABNUDA

Four days passed. I could have left the George IV at any time. I could have found cleaner, brighter, warmer, nicer, fancier accommodations, but I did not. I liked the sound of the bells. I think they came from the Church of the Holy Sepulchre. I could imagine myself in a small English village. I got used to the racy smells of the Old City, a mixture of coffee, cardamom, and donkey dung in a precise proportion. With a little prodding, I was able to get Mrs. K. Denisoglu to waddle up from her stove with a cup of tea, but I did not trust her fried food.

Around five-thirty in the afternoon there was a knock on my door. I was busy polishing up a brass astrolabe I had picked up in a shop. I opened the door. It was the boy whom Jack had hired at the Jaffa Gate to help with my bags. He came in the room and pulled a note out of his pocket and handed it to me. "Dear Professor," the note read, "I have struck oil with the Pafnuty thing. Follow Mohammed. He will lead you to the solution. Believe me, this deserves better than B plus. Jack."

I looked at the boy and pointing at him asked, "You are Mohammed?" He nodded his head and said, "You come."

It was dark out now. Many of the metal store shutters were already in place. The boy knew the Quarter the way a cat knows its own territory. He led me through alleyways and crevasses, through narrow archways and conduits, and across open ditches. At one point we crossed the Via Dolorosa and I managed to catch sight, out of the corner of my eye, of a sign in French: 7th Station of the Cross. Back into the penumbra; to the rear of a small shop where a craftsman sat at a small lathe turning out olivewood camels. More alleys. I began to wonder whether the boy was lost or misleading me.

Ultimately we came to a pastry shop. One metal shutter was up; the other window was stuffed with trays of confections and pastry goods. The boy led me through the shop and through the bakery where men were at work over kneading trays and ovens. Bags of flour were stacked on the floor. A back stair led to an upper story room. He motioned to me to follow him. We brushed past a curtain of swinging beads and I found myself in a small restaurant at the peak of its trade.

Recall, if you can, a similar scene from a Peter Lorre movie and consider that what you saw on the screen was like a Howard Johnson's restaurant compared to where I now was. The room was small and hot. It was brightly lighted with naked bulbs. It was noisy. There were about twenty small tables, each one numbered, most of them occupied. The only patrons were men in dark business suits quite suitable in Providence, Rhode Island for attending funerals. A number were wearing the kaffiyeh. Several of the patrons were already eating supper. Water pipes were puffing away. At one table, four men played cards. At another table someone read a paper. A tall waiter in a red fez and apron hopped around the tables. A charcoal stove burned at one end of the room.

The boy spotted an empty table, No. 6. He led me across the room and motioned for me to sit down. He, himself, sat down on the floor. There was a dogeared menu lying on the table. I picked it up and studied it. It was bilingual, Arabic on the right, English on the left. The English portion read

Abu Nussar
Nite Club Toples
Carte

Main Dish	Sweet	Beverage
Ox Balls	Halwua	Mitz
Kebabs	Baklawa	Tempo
Pilaf	Soumsoum	Englis Tea
		Cafe Turki

Abu Nussar's nite club did not serve hard liquor.

There was no rush. No one was going anywhere. I studied the menu. Perhaps I could teach myself Arabic by trying to transliterate. Mohammed, at my feet, was in no rush either. He had pulled out a bag of sunflower seeds and was perfectly happy.

The waiter saw me reading the menu. He came over to my table. He shook his head and pushed the menu away slightly. He said, "Jack order for you. I bring."

Well, I thought, I was all ready to try the ox balls and a cup of coffee, but if Jack has laid out a banquet in advance, so be it!

More time passed. More men came in and sat down. More sunflower seeds cracked. I thought I heard the rear of a motorcycle in the street below. Jack? I waited. No one came in.

Five minutes later the waiter took a large spoon and banged on the charcoal stove. "Nadra, Nadra," he said. The men clapped. Newspapers rustled. The card players looked up. Nadra Ibrahim flung open the beaded curtain and made her grand entrance. She was "toples."

The danse du ventre of the East is one of the great things in the world. It is like the Book of Kells or the Pythagorean Theorem. I had seen it twice in the States. Now I was to see it on native grounds, so to speak, but I would also see it performed topless in dungarees and yashmak. This was not canonical. What Nadra had evidently done was to park the Honda in the alley, run up the steps, and hoist up her halter so that it covered her lower face and neck. She pushed her belt ankh down an inch and with this quick change, she was ready to dance. I found her performance well articulated, but she did not, as Anatole France wrote of Thaïs, "enact those scenes of shame which the Pagan fables ascribe to Venus, Leda, and Pasiphae". She was not fat and forty. She might as well have been conducting a TV course in slimnastics.

When the belly dance had run its course, Nadra became a dancing waitress. This was a performance of her own invention. She didn't just haul the food over to the tables. She started from the sideboard near the stove. She loaded the orders on a little silver tray and gyrated her way across the floor and between the tables. She served the table next to mine. I could see that each order had a table number written on a slip of paper so she would know where to bring it. As she came near me, there was no sign of recognition in her eyes. She caught sight of Mohammed on the floor and spoke a sharp word to him. He got up and ran out through the curtain.

Eventually she glided over to my table with a piece of baklava and a glass of tea in a silver holder. She danced back to the stove. I looked at my sweet. Among the paper thin sheets of pastry saturated in oil and honey and raisins and nuts, I found the identifying slip. I read it:

Pa-P(e)-Nut-(Ius)
Belongs to God
Table 17.

Here was rock bottom, I thought. Beyond this one cannot go. The dancer served out all the orders. Then, unburdened by dishes, she did a reprise on the belly dance.

I watched and ate and drank and thought. Where was I? Where was Jack? How would I get back to the George IV now that Mohammed had scooted? Could I ask Nadra?

At this point I should go into a long discussion on what a fine dessert baklava is. Or perhaps I should discourse on how the dances of the East help keep both body and mind supple, or what is the best way of navigating a motorcycle through the Old City of Jerusalem after the lamps have been lit. But you probably know these things only too well.

What I suspect you don't know is that the man reading the paper at Table No. 17 was found dead the next morning in Abu Nussar's kitchen. His name was Babnuda Ismaïli. This was a false name, found on a pass that allowed him to enter at the Allenby Bridge. His real name was Atalah ibn Hassan. He was poisoned. The story was on the third page of the Jerusalem Post. You can read about it in English.

XV.

THE MAN ON THE MOON

The ocean waters spread over the face of the earth. The waters flow upwards and their continuation forms the heavenly effluvium. The goddess of these waters and of the nocturnal sky is Nut. Nut is a dark woman, and bends over the earth. Her arms and legs and breasts and wings hang vertically. Her body, studded with stars, spans the horizontal sky. Nut is the goddess of the underworld sky where the firmament hangs upside down permanently and whence by night it rises from the waters and changes place with the bright sky of the overworld.

Nut, the goddess of the sky mated with Geb, the god of the earth. Their first-born were twins Isis and Osiris, who fell in love while still in Nut's womb.

Isis: the oldest of the old, the goddess from whom all beginning arose. Isis, the great lady, the mistress of shelter, the mistress of heaven, the mistress of the House of Life, the mistress of the two Egypts. Isis, the unique, the wiser magician and more excellent than all the other gods. Isis, who was worshipped well into the year 600, longer than Buddha, longer than Christ.

Osiris: her brother and husband. In the Greek period known as Serapis, under whose patronage mathematics flourished in Alexandria.

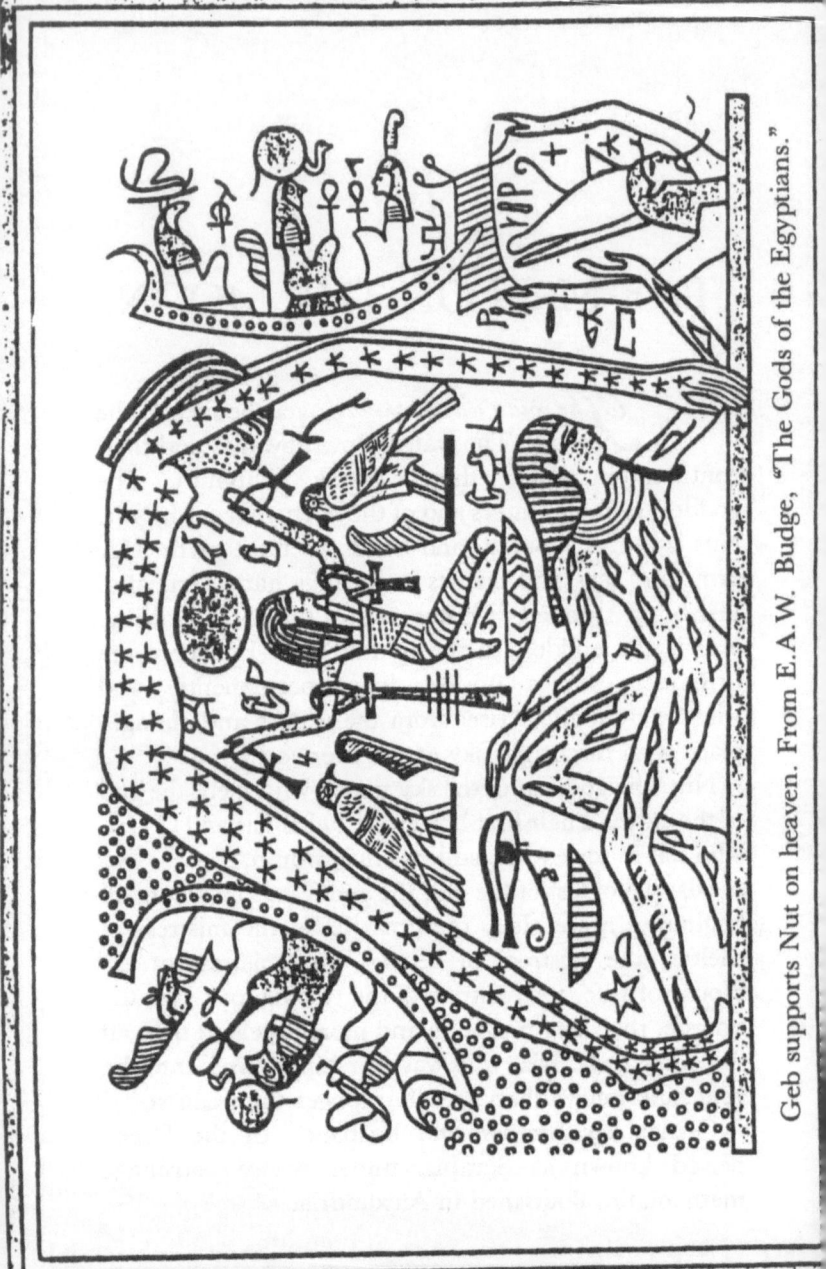

Geb supports Nut on heaven. From E.A.W. Budge, "The Gods of the Egyptians."

The technological pagans are rioting in the streets. The Age of Enlightenment is dying. The Old Gods don't work. The New Christ will deliver the coup de grace.

I have traced a thread from Tschebyscheff to Pafnuty, from Pafnuty to Paphnutius, the man of God; from Paphnutius to Papnut. I have traced Papnut back to the heiroglyphic. I shall probe no further. It does not pay to know too much.

Well, dear reader, my story is almost over. Do you know what it says? Can you read between the lines? Is it the story of a dotty professor pursuing a hobby with the madness of a zealot? The story is a true story. The people and places are real. The dead carry their true names; some of those who live hide behind pseudonyms.

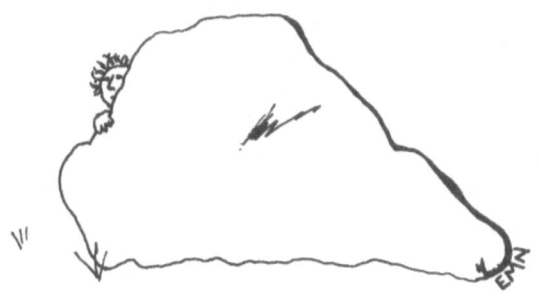

Character hiding behind a pseudonym.

What about the cross threads? What about the confluent events, the coincidences? Are they practical jokes merely, or cosmic puns as G.K. Chesterton once suggested?

Perhaps this is the story of an intelligence action: The Pafnuty Caper. Read it again. You know how to fill in details. I do not have to spell out everything. Review the events in the Mideast between 1967-1970. How do they fit? Perhaps it is the story of how one of the cleverest rings of forgers of antiquities was broken up.

I have followed a thread that led me to Russia, to Byzantium, to Egypt, to Ireland, to Sikkim, to Tasmania, to Israel. Where have I not followed it? To the moon, you say, to the moon. Very well, I shall follow it to the moon.

I have already told how Tschebyscheff was the major figure in Russian mathematics in the 1870s. He had numerous students, and there grew up a school of mathematics around him and his students that has come to be called the St. Petersburg School. Among the most distinguished members of this school we should list Korkin, Zolotareff, Markoff, Voronoi, Lyapounoff, Stekloff, Kryloff, Bernstein, and Vinogradoff. But the excellence of Russian mathematics was not preempted by the St. Petersburg School. Since Tschebyscheff's time, the Moscow School has been of increasing importance. This excellence has led the field of mathematics to "capture," in a sense, the Academy. In the early '60's, Mstislav Keldysh, a mathematician, an Approximator with strong scientific roots in Tschebyscheffiana, was named President of the Academy. This position carries far more prestige in Russia than the corresponding position in the U.S.A.

Going to the moon is an intimate marriage between technology and mathematics. Without the hardware, the propulsion, the two-way radio communication and control one has nothing. Without the differential equations of the orbits and the trajectories, without the computers performing continual predictions in real time, one also has nothing. The whole subject of space travel has a strong mathematical flavor and appeal. Sir Isaac Newton started it. He put a picture in one of his books of how a cannon ball would orbit around the earth if expelled from the cannon muzzle with a sufficiently high velocity. It's an easy consequence of his laws of motion.

A half dozen years ago, Professor Norman Levinson of M.I.T., and an international authority on the solution of differential equations, wrote a letter to the *New York Times* in which he suggested somewhat fancifully that the reason we have gone to the moon is that in the 1950's the Russian scientific establishment was in the hands of the mathematicians. There is certainly a grain of truth in this. Let me push it three generations back and say it was all due to Tschebyscheff.

Scientists have been memorialized on the moon. There are millions of natural features to be named. A good many were named in the early days of the telescope. Lunar navigation has thrown open the naming of the "dark" side as well as the smaller features of the side that faces us. There are plenty of mathematicians on the moon. Newton is there, though I think he should have been honored with a more prominent feature.

Gauss and Cauchy and Riemann are there. Tscheby-scheff is there, trust the Russians to that. His approximate coordinates are: Longitude 135° West, Latitude 30° South. He is located not too far from Norbert Wiener who taught me harmonic analysis and who was, among other things, one of the most remarkable approximators of his generation.

Shortly after the first moon walk, one of my sons got a lunar globe for his birthday. A beautiful thing. I looked for Tschebyscheff. I found him, but I was rather distressed, because his name was spelled Chebycher. By this time I was well trained by John Begg of St. Ida's, and I whipped off a letter to the manufacturer.

Rand and McNally, Division of New Moon Names: In your otherwise exemplary lunar globe, you have misspelled the name of the great Russian mathematician Pafnuty L'vovitch Tschebyscheff to whom our present moon junkets may be ascribed. Your spelling Chebysher is erroneous. I do not care one whit what system of transliteration you use, but for God's sake get the final "r" out of there. Yours faithfully.

Their story completed, their hero having come through his trial and settled down to a life of serenity and coupon clipping, Victorian novelists always provided their readers with the Grand Wrapup, projecting the lives of all their characters forward by twenty-five years. I cannot do this. I am not oracular. The best I can do is to display the ragged edges of my tapestry.

There is the story of St. Pafnuty the Tartar ascetic who founded a monastery in Borovsk, not far from where Tschebyscheff was born.

Then there is the investigation by the Israeli police of the murder of Babnuda Ismaïli and what they found.

There is the story of the time Tschebyscheff came to a meeting of French scientists in 1878 and presented a mathematical theory of how cloth can be draped on a human figures without wrinkles. I know how this story relates to Thaïs.

I know the story of the first motion picture of Thaïs with Mary Garden as Thaïs and Hamilton Revelle as Paphnutius. I know what happened at the opening.

I know what happened to Jack who deserved better than a B plus.

I know what happened to Nadra the beautiful who wore an ankh.

But, as Scheherezade said the Schahriah the Sultan, there are many nights ahead. One must leave a few threads to follow.

EPILOGUE

Ten years have passed since the previous chapters were written, and I can tell a few things that happened to the principal characters.

Professor John Begg, the cause of the whole story, has retired. His collected scientific works have appeared in two volumes.

Wilbour Hall has received a new coat of paint. Its inhabitants, though older, are all thriving. From time to time they lead me to such fine fellows as Pseudo-Aristotle or Hermes Trismegistus, the god who was real until the 16th century scholar Isaac Casaubon knocked him off.

The Met revived Thaïs with Beverly Sills in the lead role. When I read the announcement I swore that I would go, come hell or high water. The Spirits must have got wind of this vow, for two days before the performance, they dumped the heaviest blizzard ever recorded on Providence. Some people were holed up for a week. With persistence and fortitude I got to New York. The music was great, the performers were great, and to top it off for me, Beverly Sills read the unpublished version of this book. The critics did not like the production. In a faithless age, Thaïs is hard to take.

Sam married a New England woman and they are back in Sikkim running a school. Sikkim itself has fallen victim to the Zeitgeist. Caught in the struggle between India and China, it lost what little independence it had. Hope Cooke, the American Queen of Sikkim, is in New York and has written a poignant autobiography.

Jack also lives in New York. He works for a mail order firm.

Nadra immigrated to the U.S. and, surprisingly, lives not too far away. I see her from time to time. She has four children.

What about the Australian connection? I can report a few things. In the Spring of 1980, I received an invitation to address a conference of mathematicians the following January in Hobart.

Hobart in January is without a doubt the most delightful city on the face of the Earth. It is summertime. The trees are laden with apples, with plums and apricots. The lavender is in full bloom; the red hot poker plant is incredible. The temperature of the air is moderated by the cold waters of the South Pacific.

A cocktail party was announced in the name of the Vice-Chancellor of the University. The V.-C., some sort of scientist, was not around to preside at his own party. He was off to the Antarctic, looking into the penguins and the currents.

When I got to the party, people were already working on their second drink. I got mine — an innocent local apple cider that kicks like a mule after you've had six — and I wondered who or whom I should talk to. There is really no communication problem in Tasmania as English is the native language of the Island, the natives having been driven off some years ago.

I spotted a man at the far corner of the room. He looked vaguely familiar. He was drinking by himself. I went over to him and said, "You look vaguely familiar. Who are you?"

He answered, "I should be familiar, I met you here eleven years ago. My name is William Collins."

It came back to me slowly. The luncheon at the Yacht Club. Smythe-Jones. Collins. The unfortunate allusion to the spelling of Tschebyscheff. I decided not to bring up this item. I would bury the hatchet.

BIBLIOGRAPHY

V. E. Prudnikoff, "P.L. Tschebyscheff," Moscow, 1964.

B. N. Delaunay, "P.L. Chebichev and the Russian School in Mathematics," Academy Press, Moscow, Leningrad, 1945.

A. Talbot, "Approximation Theory," Inaugural Lecture. University of Lancaster, July 1971.

O. R. Kuehne, "A Study of the Thaïs Legend," Philadelphia, Pa., 1922.

A. L. Haight, ed., "Hroswitha of Gandersheim," Hroswitha Club, New York, 1965, Stechert-Hafner.

C. W. Jones, ed., Medieval Literature in Translation, Longmans, Green.

Helen Waddell, "The Desert Fathers," Constable, London, 1936.

I. Smolitsch, "Russisches Mönchtum," Würzburg, 1953.

W. E. H. Lecky, "Rationalism in Europe," Appleton, New York, 1889.

Brown University Alumni Monthly, January 1972, pp. 13-19.

Erinnerungsblätter der Mathematischen Gesellschaft zu Jena. Jena, 1859-1877.

I. V. Kuznetzova, Red., Lyudi Russkoi Nauki, Moscow, 1961.

Socrates Scholasticus, "Ecclesiastical History," Samuel
   Bagster, London, 1844.
Dov Katz, 'T'nuat Ha Musar," Tel Aviv, 5706, vol. I,
   pp. 203-204.
James Harding, "Massenet," St. Martin's Press: New
   York, 1970.
Thomas R. Hazard, "The Jonny-Cake Papers," Printed
   for the Subscribers, Boston, 1915. Reprinted: John-
   son reprints.
Henry W. Longfellow, "Tales of a Wayside Inn," Re-
   printed: David McKay Co., New York, 1966.
René de Nebesky-Wojkowitz, "Oracles and Demons of
   Tibet," Oxford University Press: London, 1956.